Response of Buried Pipelines Subject to Earthquake Effects

RESPONSE OF BURIED PIPELINES SUBJECT TO EARTHQUAKE EFFECTS

by Michael J. O'Rourke
Xuejie Liu

MULTIDISCIPLINARY CENTER FOR EARTHQUAKE ENGINEERING RESEARCH

A National Center of Excellence in Advanced Technology Applications

This monograph was prepared by the Multidisciplinary Center for Earthquake Engineering Research (MCEER) through grants from the National Science Foundation, the State of New York, the Federal Emergency Management Agency, and other sponsors. Neither MCEER, associates of MCEER, its sponsors, nor any person acting on their behalf:

    a.   makes any warranty, express or implied, with respect to the use of any information, apparatus, method, or process disclosed in this report or that such use may not infringe upon privately owned rights; or

    b.   assumes any liabilities of whatsoever kind with respect to the use of, or the damage resulting from the use of, any information, apparatus, method, or process disclosed in this report.

Any opinions, findings, and conclusions or recommendations expressed in this publication are those of the author(s) and do not necessarily reflect the views of MCEER, the National Science Foundation, Federal Emergency Management Agency, or other sponsors.

Information pertaining to copyright ownership can be obtained from the authors.

Published by the Multidisciplinary Center for Earthquake Engineering Research

    University at Buffalo
    Red Jacket Quadrangle
    Buffalo, NY 14261
    Phone: (716) 645-3391
    Fax: (716) 645-3399
    email: mceer@acsu.buffalo.edu
    world wide web: http://mceer.eng.buffalo.edu

ISBN 0-9656682-3-1

Printed in the United States of America.

Jane Stoyle, Managing Editor
Hector Velasco, Illustration
Jennifer Caruana, Layout and Composition
Jenna Tyson, Layout and Composition
Heather Kabza, Cover Design
Anna J. Kolberg, Page Design

Cover photographs provided by M. O'Rourke and T. O'Rourke.

MCEER Monograph No. 3

F O R E W O R D

Earthquakes are potentially devastating natural events which threaten lives, destroy property, and disrupt life-sustaining services and societal functions. In 1986, the National Science Foundation established the National Center for Earthquake Engineering Research to carry out systems integrated research to mitigate earthquake hazards in vulnerable communities and to enhance implementation efforts through technology transfer, outreach, and education. Since that time, our Center has engaged in a wide variety of multidisciplinary studies to develop solutions to the complex array of problems associated with the development of earthquake-resistant communities.

Our series of monographs is a step toward meeting this formidable challenge. Over the past 12 years, we have investigated how buildings and their nonstructural components, lifelines, and highway structures behave and are affected by earthquakes, how damage to these structures impacts society, and how these damages can be mitigated through innovative means. Our researchers have joined together to share their expertise in seismology, geotechnical engineering, structural engineering, risk and reliability, protective systems, and social and economic systems to begin to define and delineate the best methods to mitigate the losses caused by these natural events.

Each monograph describes these research efforts in detail. Each is meant to be read by a wide variety of stakeholders, including academicians, engineers, government officials, insurance and financial experts, and others who are involved in developing earthquake loss mitigation measures. They supplement the Center's technical report series by broadening the topics studied.

As we begin our next phase of research as the Multidisciplinary Center for Earthquake Engineering Research, we intend to focus our efforts on applying advanced technologies to quantifying building and lifeline performance through the estimation of expected losses; developing cost-effective, performance-based rehabilitation technologies; and improving response and recovery through strategic planning and crisis management. These subjects are expected to result in a new monograph series in the future.

I would like to take this opportunity to thank the National Science Foundation, the State of New York, the State University of New York at

Buffalo, and our institutional and industrial affiliates for their continued support and involvement with the Center. I thank all the authors who contributed their time and talents to conducting the research portrayed in the monograph series and for their commitment to furthering our common goals. I would also like to thank the peer reviewers of each monograph for their comments and constructive advice.

It is my hope that this monograph series will serve as an important tool toward making research results more accessible to those who are in a position to implement them, thus furthering our goal to reduce loss of life and protect property from the damage caused by earthquakes.

GEORGE C. LEE
DIRECTOR, MULTIDISCIPLINARY CENTER
FOR EARTHQUAKE ENGINEERING RESEARCH

C O N T E N T S

Foreword v
Preface xi
Acknowledgments xv
Abbreviations xvii
Notations xix

**1   Seismic Hazards and Pipeline Performance in
       Past Earthquakes** 1
1.1   Seismic Hazards 1
1.2   Performance in Past Earthquakes 2
1.3   Empirical Damage Relations 3
      1.3.1   Wave Propagation Damage 3
      1.3.2   PGD Damage 6
1.4   System Performance 11

2   Permanent Ground Deformation Hazards 13
2.1   Fault 13
2.2   Landslide 16
2.3   Lateral Spreading 20
      2.3.1   Amount of PGD 21
      2.3.2   Spatial Extent of Lateral Spread Zone 25
      2.3.3   PGD Pattern 27
2.4   Seismic Settlement 30

3   Wave Propagation Hazards 33
3.1   Wave Propagation Fundamentals 33
3.2   Attentuation Relations 35
3.3   Effective Propagation Velocity 38
      3.3.1   Body Waves 38
      3.3.2   Surface Waves 39
3.4   Wavelength 42
3.5   Ground Strain and Curvature Due to Wave Propagation 44
3.6   Effects of Variable Subsurface Conditions 46

	3.6.1	Numerical Models	47
	3.6.2	Simplified Model	54
	3.6.3	Comparison	55

4   **Pipe Failure Modes and Failure Criterion** **59**
4.1   Continuous Pipeline 59
    4.1.1   Tensile Failure 59
    4.1.2   Local Buckling 61
    4.1.3   Beam Buckling 62
    4.1.4   Welded Slip Joints 67
4.2   Segmented Pipeline 69
    4.2.1   Axial Pull-out 71
    4.2.2   Crushing of Bell and Spigot Joints 72
    4.2.3   Circumferential Flexural Failure and Joint Rotation 73

5   **Soil-Pipe Interaction** **77**
5.1   Competent Non-Liquefied Soil 77
    5.1.1   Longitudinal Movement 79
    5.1.2   Horizontal Transverse Movement 80
    5.1.3   Vertical Transverse Movement, Upward Direction 81
    5.1.4   Vertical Transverse Movement, Downward Direction 83
5.2   Equivalent Stiffness of Soil Springs 84
    5.2.1   Axial Movement 85
    5.2.2   Lateral Movement in the Horizontal Plane 85
    5.2.3   Vertical Movement 87
5.3   Liquefied Soil 87

6   **Response of Continuous Pipelines to**
    Longitudinal PGD **91**
6.1   Elastic Pipe Model 92
6.2   Inelastic Pipe Model 97
    6.2.1   Wrinkling 100
    6.2.2   Tensile Failure 101
6.3   Influence of Expansion Joints 102
6.4   Influence of an Elbow or Bend 106

7   **Response of Continuous Pipelines to**
    Transverse PGD **113**
7.1   Idealization of Spatially Distributed Transverse PGD 115
7.2   Pipeline Surrounded by Non-Liquefied Soil 116
    7.2.1   Finite Element Methods 117
    7.2.2   Analytical Methods 130
    7.2.3   Comparison Among Approaches 137
    7.2.4   Comparison with Case Histories 139
    7.2.5   Expected Response 140

7.3	Pipelines in Liquefied Soil	141
	7.3.1 Horizontal Movement	142
	7.3.2 Vertical Movement	143
7.4	Localized Abrupt PGD	147

8   Response of Continuous Pipelines to Faulting 149

8.1	Strike-slip Fault	150
	8.1.1 Analytical Models	150
	8.1.2 Finite Element Models	157
	8.1.3 Comparison Among Approaches	162
	8.1.4 Comparison with Case Histories	163
8.2	Normal and Reverse Fault	166

9   Response of Segmented Pipelines to PGD 167

9.1	Longitudinal PGD	168
	9.1.1 Distributed Deformation	168
	9.1.2 Abrupt Deformation	170
9.2	Transverse PGD	171
	9.2.1 Spatially Distributed PGD	171
	9.2.2 Fault Offsets	174

10   Response of Buried Continuous Pipelines to Wave Propagation 179

10.1	Straight Continuous Pipelines	179
	10.1.1 Newmark Approach	180
	10.1.2 Sakurai and Takahashi Approach	181
	10.1.3 Shinozuka and Koike Approach	182
	10.1.4 M. O'Rourke and El Hmadi Approach	183
	10.1.5 Comparison Among Approaches	188
	10.1.6 Comparison with Case Histories	189
10.2	Bends and Tees	191
	10.2.1 Shah and Chu Approach	191
	10.2.2 Shinozuka and Koike Approach	194
	10.2.3 Finite Element Approach	195
	10.2.4 Comparison Among Approaches	196

11   Response of Segmented Pipelines to Wave Propagation 199

11.1	Straight Pipelines/Tension	199
11.2	Straight Pipelines/Compression	204
11.3	Elbows and Connections	207
11.4	Comparison Among Approaches	209
11.5	Effects of Liquefied Soil	211

**12   Countermeasures to Mitigate
       Seismic Damage** **215**
12.1   Routing and Relocation 215
12.2   Isolation from Damaging Ground Movement 216
12.3   Reduction of Ground Movements 217
12.4   High Strength Materials 218
12.5   Flexible Materials and Joints 219

References **223**
Author Index **237**
Subject Index **241**
Contributors **249**

P R E F A C E

Buried pipeline systems are commonly used to transport water, sewage, oil, natural gas and other materials. In the conterminous United States, there are about 77,109 km (47,924 miles) of crude oil pipelines, 85,461 km (53,114 miles) of refined oil pipelines and 67,898 km (42,199 miles) of natural gas pipelines (FEMA, 1991). The total length of water and sewage pipelines is not readily available. These pipelines are often referred to as "lifelines" since they carry materials essential to the support of life and maintenance of property. Pipelines can be categorized as either continuous or segmented. Steel pipelines with welded joints are considered to be continuous while segmented pipelines include cast iron pipe with caulked or rubber gasketed joints, ductile iron pipe with rubber gasketed joints, concrete pipe, asbestos cement pipe, etc.

The earthquake safety of buried pipelines has attracted a great deal of attention in recent years. Important characteristics of buried pipelines are that they generally cover large areas and are subject to a variety of geotectonic hazards. Another characteristic of buried pipelines, which distinguishes them from above-ground structures and facilities, is that the relative movement of the pipes with respect to the surrounding soil is generally small and the inertia forces due to the weight of the pipeline and its contents are relatively unimportant. Buried pipelines can be damaged either by permanent movements of ground (i.e. PGD) or by transient seismic wave propagation.

Permanent ground movements include surface faulting, lateral spreading due to liquefaction, and landsliding. Although PGD hazards are usually limited to small regions within the pipeline network, their potential for damage is very high since they impose large deformation on pipelines. On the other hand, the wave propagation hazards typically affect the whole pipeline network, but with lower damage rates (i.e., lower pipe breaks and leaks per unit length of pipe). For example, during the 1906 San Francisco earthquake, the zones of lateral spreading accounted for only 5% of the built-up area affected by strong ground shaking. However, approximately 52% of all pipeline breaks occurred within one city block of these zones, according to T. O'Rourke et al., (1985). Presumably the remaining 48% of pipeline damage was attributed to wave propagation. Hence, although the total amount of damage due to PGD and wave propagation was roughly equal, the damage rate in the small isolated areas subject to PGD was about 20 times higher than that due to wave propagation.

Continuous pipelines may rupture in tension or buckle in compression. Observed seismic failure for segmented pipelines, particularly large diameters and relatively thick walls, is mainly due to distress at the pipeline joints (axial pull-out in tension, crushing of bell and spigot in compression). For smaller diameter segmented pipes, circumferential flexural failures (round cracks) have also been observed in areas of ground curvature.

This monograph reviews the behavior of buried pipeline components subject to permanent ground deformation and wave propagation hazards, as well as existing methods to quantify the response. To the extent possible and where appropriate, the review focuses on simplified procedures which can be directly used in the seismic analysis and design of buried pipeline components. System behavior of a buried pipeline network is not discussed in any great detail. Where alternate approaches for analysis or design are available, attempts are made to compare the results from the different procedures. Finally, we attempt to benchmark the usefulness and relative accuracy of various approaches through comparison with available case histories.

This monograph is divided into twelve chapters. Chapter 1 reviews seismic hazards and the performance of buried pipelines in past earthquakes. Chapter 2 describes the different forms of permanent ground deformation (surface faulting, lateral spreading, landsliding), and presents procedures to quantify and model both the amount of PGD as well as the spatial extent of the PGD zone. Chapter 3 reviews seismic wave propagation and presents procedures for estimating ground strain and curvature due to travelling wave effects. Chapter 4 presents the failure modes and corresponding failure criteria for buried pipelines subject to seismic effects. Chapter 5 reviews commonly used techniques to model the soil-pipe interaction in both the longitudinal and transverse directions.

Chapters 6 and 7 present the response of continuous pipelines subject to longitudinal PGD and transverse PGD respectively, while Chapter 8 discusses pipe response due to faulting. Chapter 9 presents the response of segmented pipelines subject to permanent ground deformation. Chapters 10 and 11 discuss the behavior of continuous and segmented pipeline components subject to seismic wave propagation. Chapter 12 presents current countermeasures to reduce damage to pipelines during earthquakes.

ACKNOWLEDGMENTS

This state of the art monograph is one of the products resulting from the Multidisciplinary Center for Earthquake Engineering Research (MCEER), formerly the National Center for Earthquake Engineering Research (NCEER), research projects 94-3301A and 95-3301A at Rensselaer Polytechnic Institute. These projects provided partial financial support for the second author's doctoral studies. Both authors gratefully acknowledge this support.

Much of the U.S. research reviewed in this monograph was an outgrowth of NCEER projects in the lifeline area. The NCEER lifeline activity was lead by Professor M. Shinozuka, and the authors would like to thank Professor Shinozuka for his tireless leadership of that effort.

The monograph attempts to also include key results from overseas, particularly Japan. Much of the Japanese's research was presented at a series of six U.S. – Japan workshops. This workshop series was originally organized by Professor Shinozuka of the U.S. and the late Professor K. Kubo of Japan. More recently, the workshop series was organized and lead by Professors M. Hamada (Japan) and T. O'Rourke (U.S.). Hence, in addition to their significant individual technical contributions, the authors would like to acknowledge the admirable international cooperation and professional leadership of Professors Hamada, Kubo, T. O'Rourke and Shinozuka.

ABBREVIATIONS

AC	Asbestos Cement
ASCE	American Society of Civil Engineers
ATC	Applied Technology Council
AWSS	Auxiliary Water Supply System
CI	Cast Iron
Conc	Concrete Pipe
DI	Ductile Iron
EBMUD	East Bay Municipal Utility District
ECP	Prestressed Embedded Cylinder Pipe
FE	Finite Element
FS	Factor of Safety
L-waves	Love Waves
LCP	Prestressed Lined Cylinder Pipe
LSI	Liquefaction Severity Index
MMI	Modified Mercalli Intensity
NIBS	National Institute of Building Sciences
P-waves	Compressional Waves
PE	Polyethylene
PGD	Permanent Ground Deformation
PVC	Polyvinyl Chloride
R-waves	Rayleigh Waves
RCC	Reinforced Concrete Cylinder Pipe
S-waves	Shear Waves
TCLEE	Technical Council on Lifeline Earthquake Engineering
WSAWJ	Welded Steel Arc-Welded Joints
WSCJ	Welded Steel Caulked Joints
WSGWJ	Welded Steel Gas-Welded Joints

N O T A T I O N S

A	cross-section area of pipe	e	pore ratio of soil
$a(t)$	ground acceleration as a function of time	E	modulus of elasticity
		E_i	initial Young's modulus
a_c	critical acceleration	E_p	modulus of pipe after yield
A_{core}	area of the concrete core	E_s	soil modulus, $E_s = 2(1+\mu_s)G_s$
A_m	maximum ground acceleration	F	axial force in pipe
		f	frequency in Hz
a_{max}	maximum acceleration at ground surface	F_i	axial force at the ith joint
		F_{cr}	compressive force at joint
a_x	horizontal acceleration at ground surface	F_L	liquefaction intensity factor
		F_{15}	average fines contents in T_{15} (%)
a_z	vertical acceleration at ground surface	F_R	restraint strength against axial tension
C	apparent propagation velocity of seismic wave	g	acceleration due to Earth's gravity
C_H	shear wave velocity of a half space	G, G_s	shear modulus of soil
		h, H_A, H_B	thickness of layer (m)
C_L	shear wave velocity of uniform soil layer	H	depth to center-line of pipeline
C_{ph}	phase velocity of seismic wave	H_1	thickness of saturated sand layer (m)
C_s	shear wave velocity of surface soil	H_2	height of embankment (m)
		H_c	depth to top of pipe
d	closest distance to surface projection of fault plane	H_s	thickness of uniform soil layer (m)
D	pipe diameter	I_a	Arias intensity in g's
D_A, D_B	peak ground displacements	I_p, I	moment of inertia
D_{5015}	mean grain size in T_{15} (mm)	k	reduction factor depending on outer-surface characteristics and hardness of pipe
d_a	pull-out capacity of joint (axial deformation)		
d_l	lateral deformation capacity of joint	k_o	coefficient of lateral soil pressure at rest
D_m	peak ground displacement		
D_N	Newmark displacement (cm)	K_1	equivalent soil spring for disturbed soil
D_r	relative density of soil		

K_2 — equivalent soil spring for undisturbed soil

K_c — bearing capacity factor for undrained soil

K_g — soil stiffness per unit length

K_L — soil spring constant for movement in horizontal plane

K_v — soil spring constant for downward movement

K_{v1} — soil spring coefficient for small relative displacement

K_{v2} — soil spring coefficient for moderate relative displacement

K_{v3} — soil spring coefficient for relative displacement equal to or larger than y_u

L — length of PGD zone

L' — effective slippage length at bend

L_0 — pipe segment length or distance

L_a — effective unanchored length

L_{AB} — horizontal projection of inclined rock surface

L_c — length of curved portion

L_{cr} — critical length of PGD zone

L_e — pipe length in which elastic strain develops

L_{em} — embedment length defined as the length over which the constant slippage force t_u must act to induce a pipe strain ε equal to equivalent ground strain α

L_p — pipe length in which plastic strain develops

L_s — separation distance between two stations

LSI — Liquefaction Severity Index, LSI is arbitrarily truncated at 100

M — bending moment at pipe bent

M_w — earthquake magnitude

n — number of joints within PGD zone, number of sand layers, or Ramberg Osgood parameter

N_c — bearing capacity factors for horizontal strip footings for clay

N_{ch} — horizontal bearing capacity factor for clay

N_{cv} — vertical uplift factor for clay

$(N_I)_{60}$ — corrected SPT N-value

N_q — bearing capacity factors for horizontal strip footings for sand

N_{qh} — horizontal bearing capacity factor for sand

N_{qv} — vertical uplift factor for sand

N_y — bearing capacity factors for downward loading for sand

p — internal pressure (operating pressure) in pipe

p_u — maximum resistance in horizontal transverse direction

P_w — excess pore water pressure

q_u — maximum resistance in vertical transverse direction

Q — $\dfrac{3}{16}\dfrac{K_g\lambda}{AE\zeta}$

R — source distance (km) or pipe radius

r — Ramberg Osgood parameter

r' — parameter of PGD distribution

R_c — radius of curvature of pipe

r_d — stress reduction factor varying from a value of 1 at the ground surface to a value of 0.9 at a depth of about 30 ft (10 m)

R_d distance from the epicenter to site (km)

R_e Reynolds number ($\rho VD / \eta$)

R_s closest distance to seismogenic rupture or hypocentral distance

s distance between two margins of PGD zone normalized by width W

S ground slopes (%) or shear in pipe

S_1 axial force acted on bent for Element 1

s_m normalized distance from margin of PGD zone to the location corresponding to peak transverse ground displacement

S_u undrained shear strength of surrounding soil

t pipe wall thickness

T shaking period, predominant period of soil (s) or axial tension in pipe

T_{15} thickness of saturated cohesionless soils with corrected SPT value less than 15 (m)

t_u maximum resistance in horizontal axial direction

u_j joint displacement threshold

U_g, u_g ground displacement in longitudinal direction

U_p, u_p displacement of pipeline in longitudinal direction

V velocity for pipe moving in liquefied soil

V_m maximum horizontal ground velocity

W width of PGD zone

W_a length from center of PGD zone to anchor point

W_{arc} arc length of pipe within PGD zone (m)

W_{cr} critical width of liquefied zone

W_{media} weight of medium

W_{pipe} self-weight of pipe

W_s distance between pipe supports

x non-normalized distance from the margin of the PGD zone

x_u maximum elastic deformation in horizontal axial direction

Y free face ratio (%)

y lateral displacement of soil

y_1 transverse pipe displacement in PGD zone

y_2 transverse pipe displacement outside PGD zone

y_u maximum elastic deformation in horizontal transverse direction

z_u maximum elastic deformation in vertical transverse direction

α inclined angle of slope, adhesion coefficient for clay or equivalent ground strain

α_o empirical coefficient varying with S_u

β intersection angle between pipe and fault trace

β_c, β_o conversion factors

$\beta_{optimal}$ optimal orientation of pipeline

β_p pipe burial parameter

γ total unit weight

γ_{cr} critical shear strain

γ_o maximum shear strain at pipe-soil interface

$\bar{\gamma}$ effective unit weight of soil

γ_s actual incidence angle of S-wave

δ permanent displacement of ground or pipe

δ_{cr} critical displacement of ground movement

δ_f	average fault displacement	κ_g	maximum ground curvature
Δ_I	pipe displacement at bent	λ	wavelength or beam-on-elastic foundation parameter
ΔL	total elongation of pipe		
Δu	relative displacement at joint		
Δu_c	deformation capacity of a joint in compression	μ	friction coefficient
		μ_s	Poisson ratio of soil
Δu_T	deformation capacity of a joint in tension	ξ	factor which depends on width of PGD, $0.5 \le \xi \le 1$
Δu_{ult}	relative displacement for joint closure	ρ	density of liquefied soil
		σ	uniaxial tensile stress
Δx_r	joint opening due to joint rotation	σ_o	total overburden pressure on sand layer under consideration
Δx_t	joint opening due to tension		
Δx	total maximum opening at one side of a joint due to transverse PGD	σ'_o	initial effective overburden pressure on sand layer under consideration
$\Delta\Theta$	relative rotation at pipe joint	σ_{ap}	axial stress in pipe resulting from internal pressure
ε	engineering strain	σ_{comp}	compressive strength of concrete
$\bar{\varepsilon}$	average pipe strain		
ε_o	maximum ground strain at $x=L/4$	σ_{cr}	ultimate compressive stress of segments
ε_a	maximum axial strain due to the elongation of pipe	σ_{hp}	hoop stress in pipe due to internal pressure
ε_b	pipe bending strain	σ_v	total overburden pressure
ε_{ap}	upper bound for pipe axial strain	σ'_v	effective overburden pressure
ε_g	ground strain	σ_y	apparent yield stress
ε_p	pipe axial strain	τ	parameter of PGD distribution
ε_v	volumetric strain for saturated sandy soil layer		
ε_y	yield strain	τ_{ave}	average shear stress
ζ	$\sqrt[4]{K_g / (4EI)}$	τ_s	shear force at pipe-soil interface
η	coefficient of viscosity of liquefied soil	ϕ	angle of shear resistance of sand
θ_g	slope of lower boundary of liquefied layer or ground surface	Φ	principle direction of ground motion

Seismic Hazards and Pipeline Performance in Past Earthquakes

Seismic damage to buried pipelines has been observed and documented by many individuals during post-earthquake reconnaissance. These surveys provide useful information about the failure modes and failure mechanisms of buried pipelines. Moreover, empirical damage relationships, based on statistical information from a number of past events, can be used to estimate expected damage and system performance in future earthquakes. This chapter reviews seismic hazards due to both permanent ground deformation and wave propagation effects, and presents information on the performance of both segmented and continuous pipelines in past earthquakes.

1.1 Seismic Hazards

For buried pipelines, seismic hazards can be classified as being either wave propagation hazards or permanent ground deformation hazards. There have been some events where pipe damage has been due only to wave propagation. An example is damage in Mexico City occasioned by the 1985 Michoacan earthquake. More typically, pipeline damage is due to a combination of hazards. As mentioned previously, T. O'Rourke et al. (1985) noted that roughly half the pipe breaks in the 1906 San Francisco event occurred within liquefaction-induced lateral spreading zones while the other half occurred over a somewhat larger area where wave propagation was apparently the prominent hazard. That is, permanent ground deformation (PGD) damage typically occurs in isolated areas of ground failure with high damage rates while wave propagation damage occurs over much larger areas, but with lower damage rates.

Wave propagation hazards are characterized by the transient strain and curvature in the ground due to travelling wave effects. PGD (such as landslide, liquefaction induced lateral spread and seismic settlement) hazards are characterized by the amount, geometry, and spatial extent of the PGD zone. The fault-crossing PGD hazard is characterized by the permanent horizontal and vertical offset at the fault and the pipe-fault intersectional angle. More detailed information on characterization of these PGD hazards is presented in Chapter 2.

1.2 PERFORMANCE IN PAST EARTHQUAKES

The vulnerability of buried pipelines to seismic hazards has been demonstrated by the extensive damage observed during previous earthquakes. Examples of documented pipeline damage include: the 1906 San Francisco, Manson (1908); 1933 Long Beach, Wood (1933); 1952 Kern County, Steinbrugge and Moran (1954); 1964 Alaska, Hansen (1971); 1964 Niigata, Hamada and T. O'Rourke (1992); 1971 San Fernando, McCaffrey and T. O'Rourke (1983); 1976 Guatemala, Dieckgrafe (1976); 1976 Tangshan, Sun and Shien (1983); 1979 Imperial Valley, Waller and Ramanathan (1980); 1983 Coalinga, Isenberg and Escalante (1984); 1983 Nihonkai-Chubu, Hamada and T. O'Rourke (1992); 1987 Whittier Narrows, Wang (1990); 1989 Loma Prieta, T. O'Rourke and Pease (1992); 1991 Costa Rica, M. O'Rourke and Ballantyne (1992); 1993 Kushiro-Oki, Wakamatsu and Yoshida (1994); 1994 Northridge, T. O'Rourke and Palmer (1994). Presented below is a brief description of the amount and types of pipeline damage which have been observed (T. O'Rourke et al., 1985).

- In 1964, the Anchorage, Alaska earthquake caused over 200 breaks in gas pipelines and 100 breaks in water distribution pipelines at Anchorage. Gas lines within fault zones were ruptured. Most of the pipeline damage was due to landslides and ground cracking.

- The 1971 San Fernando earthquake resulted in 1,400 breaks in various piping systems. The city of San Fernando temporally lost water, gas and sewage services. Liquefaction-induced lateral spreading along the eastern and western shores of the Upper Van Norman Reservoir damaged water, gas and petroleum transmission lines.
- The 1987 Ecuador earthquake destroyed the trans-Ecuadorian pipeline (660 mm in diameter), which represented the largest single pipeline loss in history. It cost roughly $850 million in lost sales and reconstruction.

The damage briefly discussed above was due to some combination of wave propagation and PGD effects. In the following subsection, empirical relations for pipe damage due to these hazards will be discussed separately.

1.3

EMPIRICAL DAMAGE RELATIONS

Often the first step in the seismic upgrade of a pipeline system is an evaluation of the likely amounts of damage in the existing system due to potential earthquakes. For buried pipelines, empirical correlations between observed seismic damage and some measure of ground motion are typically used. In 1975, Katayama et al. developed one of the first relations, primarily for segmented cast iron pipelines, in which damage rate is plotted as a function of peak ground acceleration. This relation, shown in Figure 1.1, includes both wave propagation and PGD damage data. As shown in Figure 1.1, the damage ratio increases by a factor of 100 for a doubling of the peak ground acceleration.

1.3.1 WAVE PROPAGATION DAMAGE

It appears that Eguchi was the first to separate wave propagation damage and PGD damage. For wave propagation, Eguchi (1983) summarized pipe break rate versus Modified Mercalli Intensity (MMI) for several earthquakes in the United States, and developed fragility relations for six different pipeline materials sub-

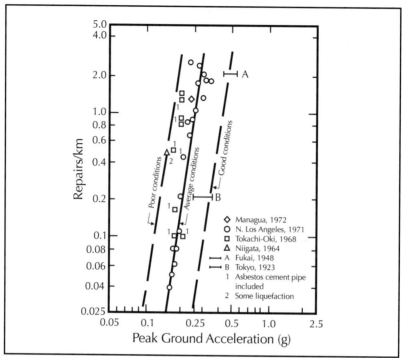

After Katayama et al., 1975

■ Figure 1.1 Pipe Damage in Repairs per Kilometer vs. Peak Ground Acceleration

ject to wave propagation only. Eguchi (1991) modified his rela-
tionship and obtained a bilinear curve shown in Figure 1.2, where
AC = Asbestos Cement, CONC = Concrete, CI = Cast Iron, PVC =
Polyvinyl Chloride, WSCJ = Welded Steel Caulked Joints, WSGWJ
= Welded Steel Gas-Welded Joints, WSAWJ (A, B) = Welded Steel
Arc-Welded Joints (Grade A, B), WSAWJ (X) = Welded Steel Arc-
Welded Joints (X Grade), DI = Ductile Iron, PE = Polyethylene.
Note that the damage rates increase fairly rapidly for MMI ≤ 8,
and then more slowly for MMI > 8.

Based on data from three U.S. earthquakes, Barenberg (1988)
established an empirical relation between seismic wave propagation damage to cast iron pipe and peak horizontal ground or particle
velocity. Note that one would expect that pipe damage to correlate fairly well with peak ground velocity since as will be shown
later ground strain, and hence pipe strain, is a function of V_{max}.
Including additional data from three other earthquakes, M.
O'Rourke and Ayala (1993) prepared a plot of wave propagation

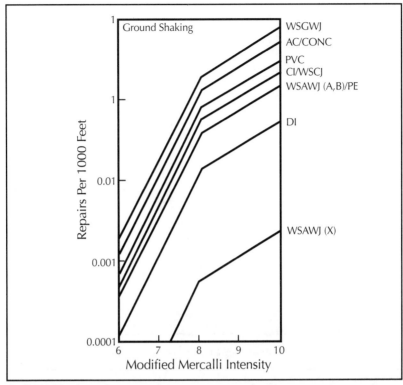

Ground Shaking

WSGWJ
AC/CONC
PVC
CI/WSCJ
WSAWJ (A,B)/PE

DI

WSAWJ (X)

Repairs Per 1000 Feet

1

0.01

0.001

0.0001

6   7   8   9   10

Modified Mercalli Intensity

After Eguchi, 1991

■ **Figure 1.2 Wave Propagation Pipe Damage vs. Modified Mercalli Intensity**

damage rate versus peak ground velocity, which includes cast-iron pipe, concrete pipe, prestressed concrete pipe and asbestos cement pipe. Both the relations are shown in Figure 1.3, where the M. O'Rourke and Ayala best straight line (point A through K) gives higher damage rates than Barenberg's. The somewhat higher estimated damage rates using the M. O'Rourke and Ayala relation were not due to the inclusion of pipe materials other than cast iron. They are thought to be due at least in part to the effects of corrosion and variable subsurface conditions.

More recently, various researchers have developed empirical wave propagation damage relations for different pipe materials (Eidinger et al., 1995) or for different diameter ranges (Honegger, 1995). Although the individual mean regression curves differ somewhat from that in Figure 1.3, the data points still fall in the general band shown in Figure 1.3 and the scatter of data points is similar.

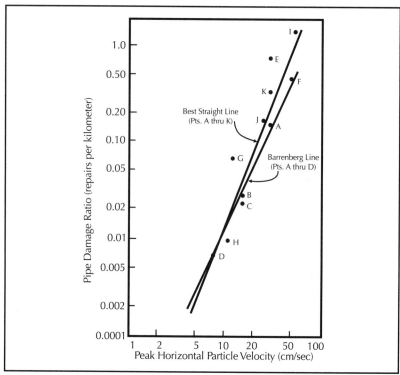

After M. O'Rourke and Ayala, 1993

■ Figure 1.3 Wave Propagation Damage to Common Water System Pipe vs. Peak Horizontal Particle Velocity

1.3.2 PGD DAMAGE

There are a variety of patterns of PGD depending on local soil conditions and the presence of faults. One type of PGD is localized abrupt relative displacement such as at the surface expression of a fault, or at the margins of a landslide. The second type of PGD is spatially distributed permanent displacement which could result, for example, from liquefaction-induced lateral spreads, or ground settlement due to soil consolidation. For localized abrupt PGD, pipeline damage mainly occurs around the ground rupture trace. On the other hand, breaks for spatially distributed PGD may occur everywhere within the PGD zone. Empirical damage relations for both types of PGD (spatially distributed and abrupt) are presented in the following discussion.

Spatially Distributed PGD

Porter et al. (1991) developed an empirical relation for bell and spigot cast iron water pipe with lead and oakum joints as shown in Figure 1.4.

As shown in Figure 1.4, the damage rate is a function of permanent ground displacement. A bilinear curve is fitted to the data from the 1906 San Francisco and 1989 Loma Prieta earthquakes. The initial portion of the curve (PGD < 5 inches (13cm)) is based on damage information for the Marina District in San Francisco during the 1989 Loma Prieta event (vertical settlements) while the later portion (PGD > 5 inches) is based upon the 1906 San Francisco event (lateral spreads).

As shown in Figure 1.4, the normalized pipe break rate is a nonlinear function of PGD. Relatively small ground displacement produces initial pipe breakage. At larger ground displacement, break rates increase, but at a smaller rate. To explain this nonlinearity, Porter et al. postulate that damage initiates at low magnitudes of PGD, breaking the original pipe network into shorter segments that are relatively free to move with the surrounding soil. Relatively larger displacements are then required to cause further breaks in the remaining intact segments.

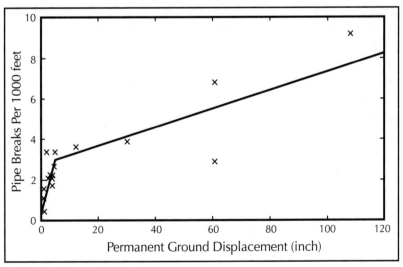

After Porter et al., 1991

■ Figure 1.4 Empirical Damage to Cast Iron Water Pipe vs. Spatially Distribution PGD

In terms of the existing empirical relation for spatially distributed PGD, Chapter 5 and 6 highlight the difference in pipe response to transverse and longitudinal PGD. In this sense, the vertical settlement data (from the 1989 Loma Prieta event) used by Porter et al. corresponds to transverse PGD. Hence, caution should be used in applying these results to situations where one expects horizontal ground movement parallel to the pipe axis (i.e., longitudinal PGD). For both types of PGD, pipe vulnerability is theoretically a function of both the amount and the spatial extent of the ground failure zone. For example, theoretical models of segmented pipe response presented in Chapter 9 suggest that a 0.5 m (18 in.) diameter CI pipe subject to 1 m (3.3 ft) of distributed transverse PGD would experience joint opening if the width of the PGD zone is 60 m (197 ft) or less, but would survive if the width is larger.

Finally, it should be noted that recent research (personal communication with Prof. T. O'Rourke) suggests that Marina District pipe damage in the 1989 Loma Prieta event was due to large transient strains resulting from shaking (wave propagation) of the "soupy" liquefied subsoil layer. Although the final (permanent) settlement, used by Porter et al., may correlate positively with the large transient strains, which apparently actually caused the damage, caution should be exercised for situation with PGD < 5 inch (13 cm). This is particularly true for settlement of dry sand where large transient strains are not expected.

For PGD damage to continuous welded steel pipes, no empirical relation determined directly from observed damage is available. However, Porter et al. developed a pseudo-empirical relation based on the damage rate of cast-iron pipelines versus permanent ground displacement, and a comparison of the damage rates for steel pipes and cast-iron pipes from Hamada and Eguchi described below.

Hamada (1989) compared cast iron pipe breakage with steel gas pipe breakage in the 1983 Nihonkai-Chubu earthquake, and concluded that the damage rate to cast iron pipes was two to three times that for steel pipes. Based on the 1971 San Fernando earthquake and the 1972 Managua earthquake, Eguchi (1983) concluded that the vulnerability of gas welded steel pipe is 30% of that for cast iron in faulting regions. In landslides, the factor is 61% and in lateral spreads it is 70%. Based on these observations,

Porter et al. assume that the rate of damage to gas welded steel pipes was half of that for cast iron pipes. Similarly, the vulnerability of arc welded steel pipes was taken as 12.5% of cast iron pipes. Porter's pseudo-empirical relations for steel pipes and various other materials are shown in Figure 1.5.

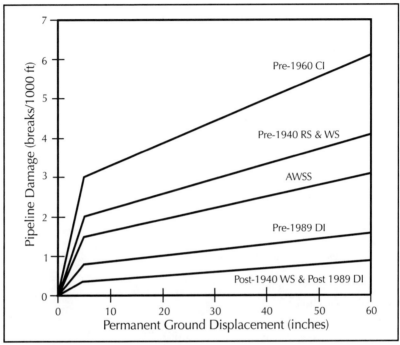

After Porter et al., 1991

■ Figure 1.5 Pipe Breaks vs. Permanent Ground Displacement

More recent studies have suggested other empirically based relations between damage and the amount of ground movement. For example, Heubach (1995) suggests that expected damage to cast iron pipe with rigid joints is roughly a factor of four larger than that for modern welded steel pipelines. Along similar lines, Eidinger et al. (1995) propose PGD damage relations which are functions of pipe material and joint type. Nevertheless, a comparison of Figures 1.1, 1.3 and 1.4 suggests that the scatter of data points about the mean regression curve is reduced when one considers wave propagation and PGD damage separately.

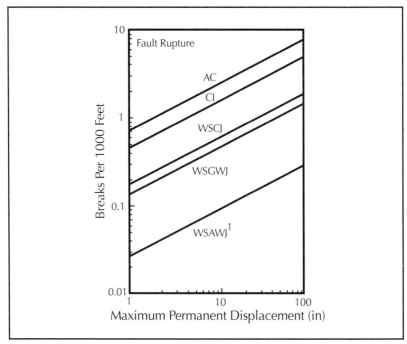

After Eguchi, 1993

■ Figure 1.6 Vulnerability Relationships for Buried Pipelines in Fault Rupture Areas

Localized Abrupt PGD

Porter et al.'s relations were obtained for spatially distributed PGD, specifically ground settlement and lateral spread. As such, it probably should not be used to predict damage to pipelines subject to localized abrupt PGD such as at a fault offset. This is because one expects higher pipe strain, for a given amount of PGD movement where the PGD is abrupt as opposed to distributed.

For fault rupture, Eguchi (1983) presented a relationship between the damage rate and the amount of fault offset, shown in Figure 1.6. It is based on damage from the 1971 San Fernando earthquake and applies to pipelines within 300 feet (91 m) each side of the predominate line of rupture. Note that the WSAWJ[1] curve presents pre-modern (i.e., prior to late 1950's) welded steel, modern (e.g. X-grade) pipe would be expected to vary by a factor of 70% lower.

As shown in Figure 1.6, the break rate per 1,000 feet for cast iron (CI) pipes is about 1.5 for abrupt PGD equal to 10 inch (25 cm) and 4.0 for abrupt PGD equal to 100 inch (2.5 m). However, for the same cast iron pipes, the break rate due to spatially distributed PGD given in Figure 1.4 is 3.2 for the displacement equal to 10 inches and 7.4 for the displacement equal to 100 inches. This result is counter-intuitive since one expects higher pipe strain for abrupt PGD. The authors understand that Eguchi currently considers the relations in Figure 1.6 to be lower bounds, with expected damage being possibly a factor of three times the lower bound value.

Even when the correct empirical relation is used, caution should be exercised in its application. As discussed in Chapter 8, the vulnerability of buried pipes to fault offset is strongly influenced by the pipe-fault intersectional angle. Since the intersectional angle is not a parameter in Eguchi's relation, the modified relation likely applies in an average sense to a wide range of intersectional angles.

The above discussion is not intended to suggest that these empirical relations are useless. They are, arguably, the best currently available. They are appropriate for evaluating overall system performance. However, by themselves, they are probably not appropriate for vulnerability analysis of an individual component.

SYSTEM PERFORMANCE

There has been a large amount of research work over the past dozen of years or so on pipeline system performance. Notable contributions have been made by Isoyama and Katayama (1982), Liu and Hou (1991), Sato and Shinozuka (1991), Honegger and Eguchi (1992), and Markov et al. (1994). A detailed discussion of overall system modeling and performance is beyond the scope of this state-of-the-art review, which focuses primarily on component performance, behavior and design. However, a summary of the results of system performance evaluations as a function of buried pipeline component performance (specifically breaks per unit length) will be discussed briefly.

Isoyama and Katayama (1982) evaluated water system performance following an earthquake for two supply strategies: supply priority to nodes with larger demands, and supply priority to nodes with lowest demands. These two strategies correspond to the best and worst system performance, which is shown in Figure 1.7. Recently, Markov et al. (1994) evaluated the performance of the San Francisco auxiliary water supply system (AWSS), while G&E (1994) did a similar study for the water supply system in the East Bay Municipal Utility District (EBMUD). Their results are also shown in Figure 1.7. Based on these results, NIBS (National Institute of Building Sciences) (1996) proposed a damage algorithm, in which the system serviceability index is a lognormal function of the average break rate. Note that in this figure, the serviceability index is considered as a measure of reduced flow.

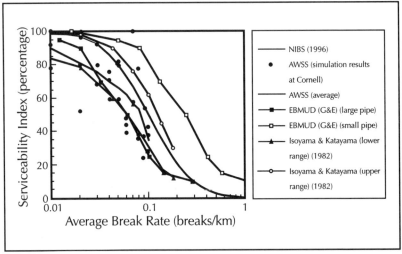

After NIBS, 1996

■ Figure 1.7 Serviceability Index vs. Average Break Rate for Post-Earthquake System Performance Evaluation

PERMANENT GROUND
DEFORMATION HAZARDS

The principal forms of permanent ground deformation (PGD) are surface faulting, landsliding, seismic settlement and lateral spreading due to soil liquefaction. Whether the buried pipeline fails when subjected to PGD depends, in part, on the amount and spatial extent of the PGD, which are introduced here.

The aim of this chapter is to provide a general overview of permanent ground deformation. First, we discuss the types of faults and the expected amount of fault offset, which is empirically correlated with earthquake magnitude. Second, we describe the types of landslides, empirical relations for the occurrence of landslides, and analytical relations for the amount of earth flow movement. Third, two approaches to evaluate ground settlement induced by soil liquefaction are introduced. Finally, we present the characteristics of lateral spreads induced by soil liquefaction.

2.1 FAULT

An active fault is a discontinuity between two portions of the earth crust along which relative movements can occur. The movement is concentrated in relatively narrow fault zones. Principal types of fault movement include strike-slip, normal-slip and reverse slip as shown in Figure 2.1. In a strike-slip fault the predominant motion is horizontal, which deforms a continuous pipe primarily in tension or compression depending on the pipe-fault intersectional angle. In normal and reverse faults the predominant ground displacement is vertical. When the overhanging side of the fault moves downwards, the fault is normal, which deforms a horizontal pipe primarily in tension. When the overhanging side of the fault moves upwards, the fault is reverse, which deforms a horizontal pipe primarily in compression.

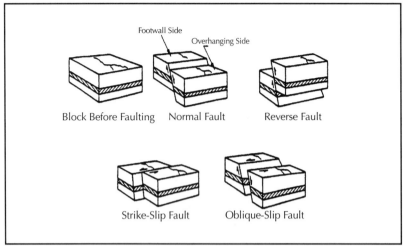

After Meyersohn, 1991

■ Figure 2.1 Block Diagrams of Surface Faulting

As mentioned previously in Section 1.2, the strain in a continuous pipe subject to fault offset depends on the amount of the fault offset and the pipe-fault intersectional angle. Here we only discuss the amount of fault offset. The effects of the intersectional angle will be discussed in Chapter 8.

Various empirical relations between fault displacement and moment magnitude have been proposed. They all have a similar logarithmic form. Here we only introduce the relationship by Wells and Coppersmith (1994) because it extends previous studies by including data from recent earthquakes and from new investigations of older earthquakes. Based on a worldwide data base of 421 historical earthquakes, Wells and Coppersmith selected 244 earthquakes, and developed the following empirical relationships:

$$\log\delta_f = -6.32 + 0.90M \text{ for Strike-Slip Fault} \qquad (2.1)$$

$$\log\delta_f = -4.45 + 0.63M \text{ for Normal Fault} \qquad (2.2)$$

$$\log\delta_f = -0.74 + 0.08M \text{ for Reverse Fault} \qquad (2.3)$$

where δ_f is the average fault displacement, in meters, M is the moment magnitude. The observed fault displacement in the Wells and Coppersmith's data base (i.e., magnitude range from 5.6 to 8.1) varies from 0.05 to 8.0 m for strike-slip faults, 0.08 to 2.1 m for normal faults, and 0.06 to 1.5 m for reverse faults as shown in Figure 2.2. The maximum fault displacement is twice the average fault displacement, according to Wells and Coppersmith. Note that the single curve (i.e., solid lines) in Figure 2.2(a) is for a combined model, while the three curves in Figure 2.2 (b) are for strike slip, reverse and normal faults respectively.

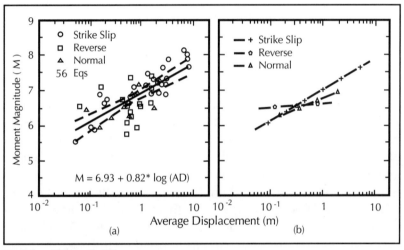

After Wells and Coppersmith, 1994

■ Figure 2.2 Regression of Average Surface Displacement on Magnitude

If a fault is poorly known or blind (i.e., lack of clear surface expression to judge fault type), the all-slip-type regression provided by Wells and Coppersmith can be used to evaluate the expected fault displacement.

$$\log\delta_f = -4.80 + 0.69M \text{ for all} \tag{2.4}$$

LANDSLIDE

Landslides are mass movements of the ground which may be triggered by seismic shaking. A large number of systems have been developed to classify landslides. The most widely used classification system in the United States was devised by Varnes (1978). Varnes identified five principal categories based on soil movements, geometry of the slide, and the types of material involved. Varnes's categories are: falls, topples, slides, spreads, and flows. Herein, lateral spreading is considered to be a liquefaction-induced phenomenon, and is discussed in Section 2.3.

Based on the different effects on pipelines, Meyersohn (1991) established three types of landslides as shown in Figure 2.3.

As shown in Figure 2.3, Type I includes rock fall and rock topple, which can cause damage to above-ground pipelines by direct impact of falling rocks. This type of landslide has relatively little effect on buried pipelines and will not be discussed in detail. Type II includes earth flow and debris flow, in which the transported material behaves as a viscous fluid. Large movements (several meters or more) are often associated with this type of landslide but the expected amount of movement is difficult to predict. Type II landslides will not be discussed herein. Type III includes earth slump and earth slide, in which the earth moves, more or less as a block. They usually develop along natural slopes, river channels, and embankments. Because pipelines often cross such zones, the following will focus on this type of landslide.

Empirical methods have been used to determine upper bounds for the occurrence of landslides. Figure 2.4 shows one such relation (Applied Technology Council, 1985), in which the maximum distance of observed landslides to the fault rupture zone is plotted as a function of earthquake magnitude.

Recent work by Jibson and Keefer (1993) resulted in analytical estimation of the expected amount of landslide movement. They used the computer program STABL (Siegel, 1978) to search for the critical failure surface by randomly generating slip surfaces and calculating the factor of safety (FS). The factor of safety is the ratio

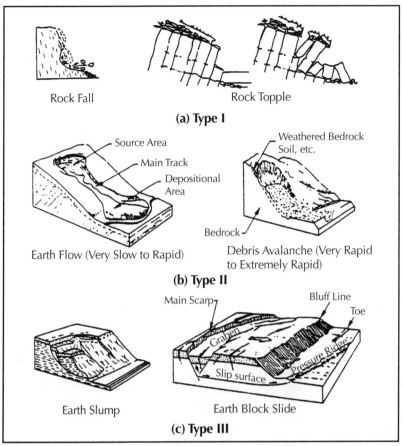

Rock Fall

Rock Topple

(a) Type I

Source Area

Main Track

Depositional Area

Weathered Bedrock Soil, etc.

Bedrock

Earth Flow (Very Slow to Rapid)

Debris Avalanche (Very Rapid to Extremely Rapid)

(b) Type II

Main Scarp

Bluff Line

Toe

Graben

Slip surface

Pressure Ridge

Earth Slump

Earth Block Slide

(c) Type III

After Meyersohn, 1991

■ **Figure 2.3 Selected Ground Failure Associated with Landsliding**

of the sum of the resisting forces and sum of the driving forces that tend to cause movement. That is, the critical failure surface is the slip surface with the lowest factor of safety.

Based on Newmark's Block model (Newmark, 1965), the critical acceleration, a_c, can then be defined as:

$$a_c = g(FS - 1)\sin\alpha \qquad (2.5)$$

where g is the acceleration due to gravity and α is the inclined angle of the slope.

The displacement of the block is then calculated by double integration of the ground accelerations larger than the critical acceleration a_c.

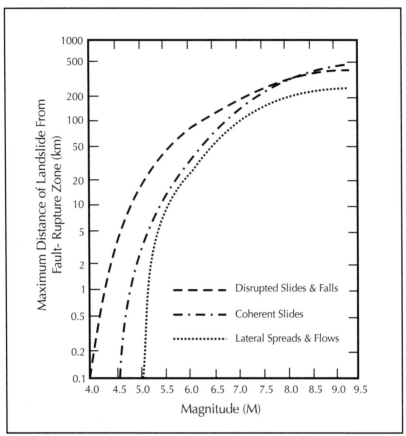

■ Figure 2.4 Occurrence of Landslide vs. Magnitude of Earthquake

Jibson and Keefer selected 11 strong-motion records to estimate the Newmark displacement. For each of the strong-motion records, they calculated the Newmark displacement for several critical accelerations between 0.02 and 0.4 g, which is considered to be the practical range of interest for most earthquake-induced landslides. The resulting data are plotted in Figure 2.5, for which the best regression function is:

$$\log D_N = 1.460 \log I_a - 6.642 a_c + 1.546 \qquad (2.6)$$

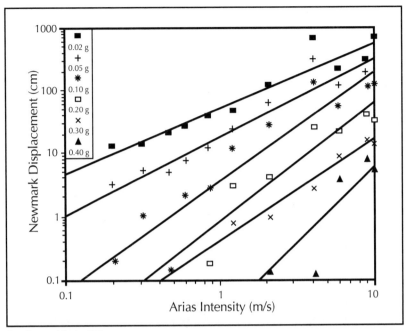

■ Figure 2.5 Newmark Displacement vs. Arias Intensity for Critical Accelerations of 0.02-0.40g

where D_N is the Newmark displacement in centimeters and I_a is the Arias intensity in g's, defined as:

$$I_a = \frac{\pi}{2g} \int [a(t)]^2 dt \qquad (2.7)$$

where $a(t)$ is the ground acceleration time history.

In this regard, Wilson and Keefer (1983) developed a simple relationship between Arias intensity, earthquake magnitude, M, and source distance, R, in kilometers:

$$\log I_a = M - 2\log R - 4.1 \qquad (2.8)$$

Note that Equation 2.8 is developed from California earthquakes and may slightly underestimate shaking intensity in the central United States.

LATERAL SPREADING

Lateral spreads develop when a loose saturated sandy soil deposit is liquefied due to seismic shaking. Liquefaction causes the soil to lose its shear strength, which in turn results in the flow or lateral movement of liquefied soil. Although the ground movement is primarily horizontal, Towhata et al. (1991) observed that vertical soil movement often accompanies liquefaction-induced lateral spreading. However, the vertical component is typically small and will be disregarded herein.

In terms of pipeline response, two situations are possible. In the first case such as at the Ogata Primary School site during the 1964 Niigata event, the top surface of the liquefied layer is essentially at the ground surface. For this first case, a pipeline is subject to horizontal force due to liquefied soil flow over and around the pipeline, as well as uplift or buoyancy forces. In the second case such as at the Mission Creek site during the 1906 San Francisco event, the top surface of the liquefied layer is located below the bottom of a typical pipeline. That is, the pipeline is contained in a non-liquefied surface soil layer which rides over the liquefied layer. For this second case, the pipeline is subject to horizontal forces due to non-liquefied soil-structure interaction but not subject to buoyancy effects. Pipeline response to such horizontal loading is discussed in Chapter 6 and 7. Pipeline response to buoyancy forces is discussed in Chapter 7.

The direction of movement for the lateral spread is controlled by geometry. When the lateral spread occurs at or near a free face, the movement is generally towards the free face. When the lateral spread occurs away from a free face, the movement is down the slope of the ground surface or down the slope of the bottom of the liquefied layer. For the "PGD towards a free face" data, the observed distance is from 10 to 300 m (33 to 984 ft) away from the free face with average value of 100 m (Bartlett and Youd, 1992). For the "PGD away from a free face" data, the observed slope is from 0.1% to 6% with an average value of 0.55%.

There are four geometric characteristics of a lateral spread which influence pipeline response in a horizontal plane. With reference to Figure 2.6, these are the amount of PGD movement δ, the transverse width of the PGD zone W, the longitudinal length of the PGD zone L, and the pattern or distribution of ground movement across and along the zone.

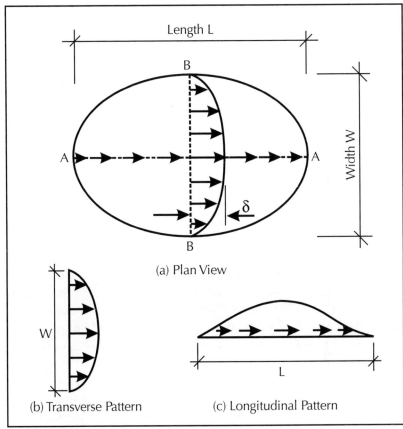

■ Figure 2.6 Characteristics of a Lateral Spread

In the following subsection, available information which can be used to quantity each of these characteristics is presented.

2.3.1 AMOUNT OF PGD

In general, the potential for PGD to induce pipe damage is related to the amount of ground movement, the length and width of the PGD zone as well as the pattern of deformation. Predicting the amount of ground displacement due to soil liquefaction is a challenging problem. Nevertheless there have been a number of studies, both analytical and empirical, which have addressed this issue. These studies are reviewed below.

Analytical and Numerical Models

Dobry and Baziar (1990), and Mabey (1992) estimate lique-faction-induced displacement using a Newmark sliding block analysis. In this analysis a 1-D, rigid soil block model is allowed to displace along a planar failure surface during time intervals when the sum of the inertial (i.e., earthquake) and gravity (i.e., self weight) components along the slide surface exceeds the shear strength of the soil.

Hamada et al. (1987), Towhata et al. (1991), and Yasuda et al. (1991) used 2-D, static elastic models to estimate the amount of lateral spreading displacement. They model the non-liquefied surface layer as a 2-D, elastic beam, floating on the liquefied layer below, and subject to lateral loading (the component of the gravity load parallel to the inclined ground surface). An analytical, closed form solution is used to calculate the ground displacement by minimizing the potential energy of the system.

Orense and Towhata (1992) used a variational principle to develop a 3-D analytical relation for the amount of liquefaction-induced ground displacement. The method is based on the principle of minimum potential energy. The lateral displacements are calculated based on the assumption on a half sinusoidal distribution of lateral displacement along a vertical section (i.e., zero at the bottom and maximum at the top) and vertical displacements are calculated based on constant volume assumption. The Rayleigh-Ritz method is employed to obtain the solution.

However, as pointed out by Bartlett and Youd (1992), these analytical and numerical models have not been applied to a wide range of earthquake and site conditions. More validation and calibration studies are likely needed before these analytical and numerical techniques can be used directly by practicing engineers.

Empirical Model

Several empirical models have been proposed to predict lateral spread displacements. The following brief review describes these existing empirical relations, their underlaying assumptions and expected range of applicability.

Work by Hamada et al. (1986) suggests that the amount of PGD induced by liquefaction is closely related to the geometric configuration of the estimated liquefied layer. They proposed the following regression formula for the magnitude of horizontal PGD, δ, in meters:

$$\delta = 0.75\sqrt{h} \cdot \sqrt[3]{\theta_g} \qquad (2.9)$$

where h is the thickness of the liquefied layers, in meters, and θ_g is the slope of the lower boundary of the liquefied layer or the ground surface (%), whichever is larger.

Note that the Hamada et al. relation does not distinguish between the amount of expected PGD at a free face as opposed to that for gently sloping ground. In addition, the thickness of the liquefied layer is in a sense a pseudo parameter which accounts for the amount of ground shaking (related to earthquake magnitude and distance) as well as the soil characteristics at the site. According to Bartlett and Youd (1992), it produces reasonable estimate for earthquakes with magnitude around 7.5 and epicenter distance in the 20 to 30 km range.

Youd and Perkins (1987) introduced the concept of a Liquefaction Severity Index (LSI) which is defined as the amount of PGD, in inches, associated with lateral spreading on gently sloping ground and poor soil conditions. LSI is arbitrarily truncated at 100. Youd and Perkins established a correlation between LSI, earthquake magnitude and distance for the western U.S. as follows:

$$\log LSI = -3.49 - 1.86\log R_d + 0.98M_w \qquad (2.10)$$

where R_d is the distance from the epicenter to the site, in kilometers for western U.S. earthquakes, and M_w is the earthquake magnitude.

Using that correlation as a starting point and data available for the 1811-12 New Madrid earthquakes, Turner and Youd (1987) proposed the following relation for the New Madrid area:

$$\log LSI = 4.252 - 1.276\log R_d \qquad (2.11)$$

Two separate equations are provided to account for differences in attenuation of strong ground motion east and west of the Rocky Mountains. Note that LSI given above is not a function of local soil parameters (i.e., applies to the worst possible soil condition) nor ground slope (it applies to ground slopes between 0.5 and 5%). In addition, as with Hamada's work, the LSI relation does not distinguish between the expected amount of PGD at a free face as opposed to that for gently sloping ground condition.

Using data from 1906 San Francisco, 1964 Alaska, 1964 Niigata, 1971 San Fernando, 1979 Imperial Valley, 1983 Nihonkai-Chubu, 1983 Borah Peak and 1987 Superstition Hills earthquakes, Bartlett and Youd (1992) recently developed two empirical relations for the expected amount of PGD due to liquefaction. The first is for lateral spreads down gentle ground slopes and the other is for lateral spreads at a free face.

For gently sloping ground condition, the relation is:

$$\log(\delta + 0.01) = -15.787 + 1.178M - 0.927\log R_d - 0.013R_d$$
$$+ 0.429\log S + 0.348\log T_{15} + 4.527\log(100 - F_{15}) - 0.922D_{50_{15}}$$

$$(2.12)$$

For PGD at a free face

$$\log(\delta + 0.01) = -15.787 + 1.178M - 0.927\log R_d - 0.013R_d$$
$$+ 0.429\log Y + 0.348\log T_{15} + 4.527\log(100 - F_{15}) - 0.922D_{50_{15}}$$

$$(2.13)$$

where δ is the permanent horizontal displacement of ground (m), M is the earthquake magnitude, R_d is the epicentral distance (km), S is the ground slopes (in percent, shown in Figure 2.7(a)), Y is the free face ratio (in percent, shown in Figure 2.7 (b)), F_{15} is the average fines contents in T_{15} (%), $D_{50_{15}}$ is the mean grain size in T_{15} (mm) and T_{15} is the thickness of saturated cohesionless soils with a corrected SPT value less than 15, (m).

Both equations include the effects of shaking at the site, soil properties and site topography. For a given amount of ground shaking (i.e., magnitude and epicenter distance fixed), the parameters

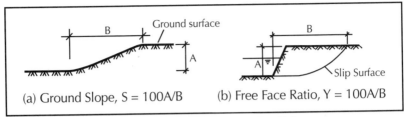

(a) Ground Slope, S = 100A/B   (b) Free Face Ratio, Y = 100A/B

■ Figure 2.7 Elevation View Showing Ground Slope and Free Face Ratio

which most strongly influence the amount of PGD are the average fines contents, followed by the mean grain size and the ground slope/free face ratio. The accuracy of the Bartlett and Youd (1992) empirical relations are relatively good, in that predicted values are generally within a factor of two of the observed values. As such, they are arguably the best currently available relations for use in the western U.S.

2.3.2 SPATIAL EXTENT OF LATERAL SPREAD ZONE

As will be seen later, the width and the length of the PGD zone have a strong influence on pipe response to PGD. Unfortunately the currently available information on the spatial extent of lateral spread zone is somewhat limited. Although one expects that the spatial extent of the lateral spread zone strongly correlates with the plan dimensions at the area which liquefied, analytical or empirical relations are not currently available. In the following, both the width W and length L as shown in Figure 2.6 will be discussed.

Information on observed values for the spatial extent of the lateral spread zone has been developed by Suzuki and Masuda (1991). Using data from the 1964 Niigata and 1983 Nihonkai-Chubu earthquakes, they presented scattergrams in Figure 2.8 of the amount of ground movement and spatial extent of the lateral spread zone for PGD away from a free face. Note that most all the observed widths are distributed in the range of about 80 to 600 m (262 to 1968 ft) and the lateral displacement tends to increase with increasing width.

In terms of the length of the lateral spread zone at a free face, the study by Bartlett and Youd (1992) provides useful information.

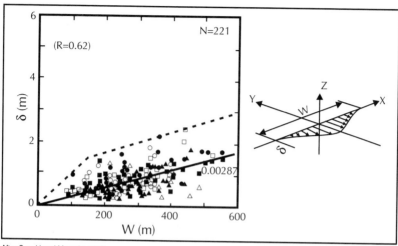

After Suzuki and Masuda, 1991

■ **Figure 2.8 Observed Data on the Amount of PGD and the Width of the Lateral Spread Zone Away From a Free Face**

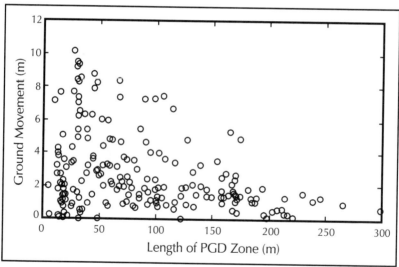

After Bartlett and Youd, 1992

■ **Figure 2.9 Observed Data on the Amount of PGD and the Length of the Lateral Spread Zone at a Free Face**

Figure 2.9 shows observed data on the amount of PGD and the length of the lateral spread zone at a free face. As with the observed PGD zone width shown in Figure 2.8, the observed lengths in Figure 2.9 are less than about 400 m (1312 ft), with most of the values below 200 m (656 ft). Although there is a large amount of scatter, the ground displacement appears to be a decreasing function of the length of the lateral spread zone for this free face situation. On the other hand, as discussed in relation to Figure 2.8, the ground displacement appears to be an increasing function of the width of the lateral spread zone for gently sloping ground situations.

Nevertheless, due to the large amount of scatter in these figures, it seems that the expected length and width of a lateral spread zone, particularly for site specific studies, should be based upon the expected plan area of liquefaction as opposed to the estimated ground movement.

2.3.3   PGD PATTERN

As noted previously, the response of buried pipelines to PGD is influenced by the pattern of deformation, that is the variation of permanent ground displacement across the width (Figure 2.6(b)) or along the length (Figure 2.6(c)) of the lateral spread zone. The study by Hamada et al. (1986) of liquefaction in the 1964 Niigata earthquake and 1983 Nihonkai-Chubu earthquake provide a wealth of information on observed longitudinal PGD patterns. Figure 2.10 shows longitudinal PGD observed along five of 27 lines in Noshiro City resulting from the 1983 Nihonkai-Chubu earthquake. In this figure the height of the vertical line is proportional to the observed horizontal PGD at the point.

Note that about 20% of the observed patterns (6 out of 27) have the same general shape as Figure 2.10(a). That is, they show relatively uniform PGD movement over the whole length of the lateral spread zone. The response of continuous buried pipeline to idealizations of these longitudinal patterns of PGD is discussed in Chapter 6.

Information on transverse patterns of PGD, as shown in Figure 2.6(b) is more limited. Figure 2.11 shows five transverse PGD patterns observed in the 1971 San Fernando earthquake and 1964 Niigata earthquake.

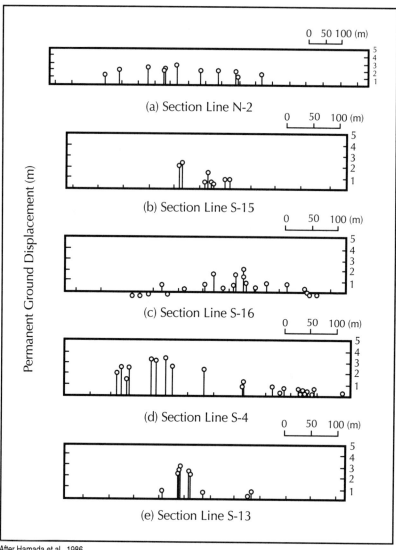

0 50 100 (m)

(a) Section Line N-2

0   50   100 (m)

(b) Section Line S-15

0   50   100 (m)

(c) Section Line S-16

0   50   100 (m)

(d) Section Line S-4

0   50   100 (m)

(e) Section Line S-13

Permanent Ground Displacement (m)

After Hamada et al., 1986

■ Figure 2.10 Observed Ground Deformation

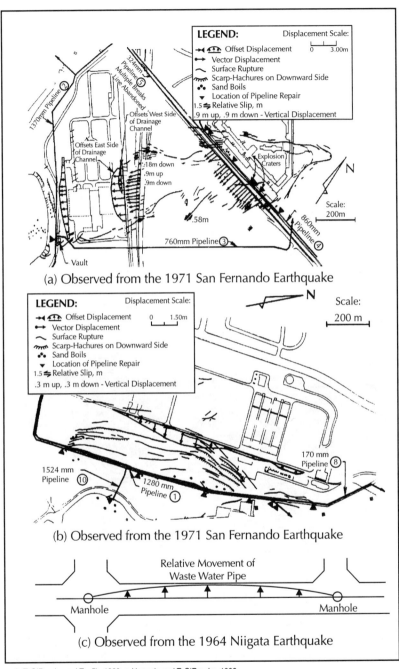

(a) Observed from the 1971 San Fernando Earthquake

(b) Observed from the 1971 San Fernando Earthquake

(c) Observed from the 1964 Niigata Earthquake

a., b. T. O'Rourke and Tawfik, 1983; c. Hamada and T. O'Rourke, 1992

■ Figure 2.11 Observed Transverse PGD Patterns

SEISMIC SETTLEMENT

Earthquake induced subsidence may be caused by densification of dry sand, consolidation of clay or consolidation of liquefied soil. Among these three types, the liquefaction-induced ground settlement is somewhat more important in that it can lead to larger ground movement and hence higher potential for damage to buried pipeline system. Ground settlement induced by soil liquefaction is discussed below.

An example of observed seismic settlement due to liquefaction is shown in Figure 2.12. This figure presents contours of ground settlement in the Marina District occasioned by the 1989 Loma Prieta earthquake (T. O'Rourke et al., 1991).

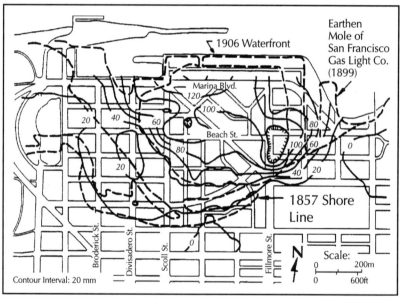

After T. O'Rourke et al., 1991

■ Figure 2.12 Contour of Ground Settlement in Marina District

Note that the maximum ground settlement is about 140 mm (5.5 in). In comparison to the expected amount of lateral spread deformation discussed previously, the expected amount of ground settlement, for the same general level of ground shaking, are typically smaller.

For saturated sands without lateral spread movement, Tokimatsu and Seed (1987) developed an analytical procedure to evaluate ground settlement. The fundamental relation is:

$$\delta = \sum (\varepsilon_v)_i h_i \qquad i = 1, 2, ..., n \qquad (2.14)$$

where ε_v is the volumetric strain for a saturated sandy soil layer, h is the layer thickness and n is the number of sand layers with different SPT N-values.

The volumetric strain in each layer depends on the SPT N-value and the cyclic stress ratio as shown in Figure 2.13, where $(N_1)_{60}$ is the corrected SPT N-value.

The cyclic stress ratio can be computed by:

$$\frac{\tau_{ave}}{\sigma_o'} = 0.65 \frac{a_{max}}{g} \cdot \frac{\sigma_o}{\sigma_o'} \cdot r_d \qquad (2.15)$$

where a_{max} is the maximum acceleration at the ground surface, σ_o and σ_o' are the total overburden pressure and the initial effective overburden pressure on the sand layer under consideration and r_d is the stress reduction factor varying from a value of 1 at the ground surface to a value of 0.9 at a depth of about 10 m (30 ft) (Seed et al., 1987).

Note that T. O'Rourke et al. (1991) used a similar approach to estimate liquefaction induced settlement in the Marina District. As noted by T. O'Rourke et al., there is good agreement between the estimated and measured settlements for the natural soils and land-tipped fill, but the estimated settlements of the hydraulic fill are almost twice as much as those observed in the field.

Seed et al.'s analytical approach provides a relatively accurate estimate of ground settlement. However, it is somewhat complex and requires detailed information on site condition and soil properties.

Takada and Tanabe (1988) developed two empirical regression equations for liquefaction-induced ground settlement at embankments and plain (level) sites based on 404 observations during five Japanese earthquakes.

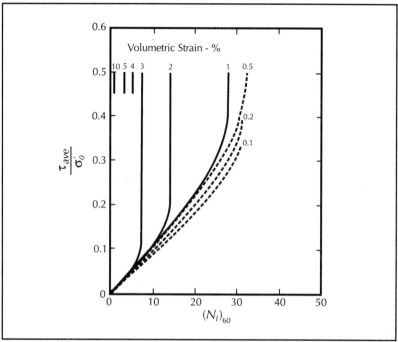

After Tokimatsu and Seed, 1987

■ **Figure 2.13 Relation between Cyclic Stress Ratio, $(N_l)_{60}$, and Volumetric Strain for Saturated Sands**

For embankment:

$$\delta = 0.11H_1H_2a_{max}/N+20.0 \qquad (2.16)$$

For plain site:

$$\delta = 0.30H_1a_{max}/N+2.0 \qquad (2.17)$$

where δ is the ground settlement in centimeters, H_1 is the thickness of saturated sand layer (in meters), H_2 is the height of embankment (in meters), N is the SPT N-value in the sandy layer, and a_{max} is the ground acceleration in gals. Takada and Tanabe's empirical approach is simple but somewhat less accurate than the Seed et al.'s approach.

WAVE PROPAGATION
HAZARDS

The wave propagation hazard for a particular site is characterized by the peak ground motion parameters (acceleration and
velocity) as well as the appropriate propagation velocity. This chapter briefly reviews attenuation relations for peak ground parameters,
as well as simplified procedures for determining the apparent propagation velocities for both body and surface waves. The ground strain
and curvature due to wave propagation are then presented. Finally, the influence of variable subsurface conditions on ground
strain is discussed.

3.1

WAVE PROPAGATION
FUNDAMENTALS

There are two types of seismic waves: body waves and surface
waves. The body waves propagate through earth, while the surface
waves travel along the ground surface. The body waves are generated by seismic faulting, while for the simplest case surface waves
are generated by the reflection and refraction of body waves at the
ground surface. Body waves include compressional waves (P-
waves) and shear waves (S-waves). In compressional waves, the
ground moves parallel to the direction of propagation, which gen-
erates alternating compressional and tensile strain. For S-waves,
the ground moves perpendicular to the direction of propagation.

The situation for surface waves is somewhat more complex.
Rayleigh and Love waves are two main types of surface waves
generated by earthquakes. For the Love waves (L-waves), the particle motion is along a horizontal line perpendicular to the direction
of propagation, while for R-waves the particle motion traces a retrograde ellipse in a vertical plane with the horizontal component

of motion being parallel to the direction of propagation. For both L- and R-waves, the amplitude of motions decreases with depth below the ground surface.

Figure 3.1 shows the East-West ground velocity histories in the hill and lake zones of Mexico City during the 1985 Michoacan earthquake. Notice in the hill zone record, the peak ground velocity is about 10 cm /s (Figure 3.1(a)) and the ground motion dies out about 60 s after initial triggering. In the lake zone record (Figure 3.1(b)), the ground velocity during the first 30 to 40 s after initial triggering is roughly about 10 cm/s. However the peak ground velocity of 30 or 40 cm/s occurs roughly a minute or two after initial triggering. This suggests that Rayleigh waves could well have been present in the lake zone. Note that if R-waves are present, they occur after the arrival of the direct body waves. That is, P-waves arrive at a site first, followed by S-waves. If surface waves are present, they typically arrive after the body waves.

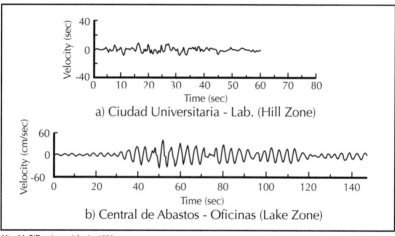

After M. O'Rourke and Ayala, 1990

■ **Figure 3.1 Ground Velocity Time History in Hill and Lake Zones During the 1985 Michoacan Earthquake**

ATTENTUATION RELATIONS

Many authors have developed empirical attenuation relations for peak ground acceleration, velocity and displacement. T. O'Rourke et al. (1985) presents a review and summary of many of the existing relations. Those relations typically are functions of earthquake magnitude, site-to-source distance and local site conditions. More recently, relations accounting for fault mechanism have been developed. This section introduces some of these attenuation relations which were based on relatively large amounts of data.

One of the most recent relations for peak ground acceleration is that proposed by Campbell and Bozorgnia (1994). They used 645 near-source accelerograms from 47 worldwide earthquakes from 1957 to 1993 to update the strong-motion attenuation relation. They found that reverse and thrust earthquakes generate larger acceleration than strike-slip earthquakes, and that soft rock has higher acceleration than hard rock. Their relation is given by:

$$\ln A_m = -3.512 + 0.904 M_w - 1.328 \ln\sqrt{R_s^2 + \left[0.149 \exp(0.647 M_w)\right]^2}$$
$$+ \left[1.125 - 0.112 \ln(R_s) - 0.0957 M_w\right] F$$
$$+ \left[0.440 - 0.171 \ln(R_s)\right] S_{sr} + \left[0.405 - 0.222 \ln(R_s)\right] S_{hr}$$

$$(3.1)$$

where A_m is the horizontal component of peak ground acceleration (g), M_W is the moment magnitude, R_s is the closest distance to the seismogenic rupture of the fault (km), $F=0$ for strike-slip and normal faulting earthquakes and 1 for reverse, reverse-oblique and thrust faulting earthquakes, $S_{sr}=1$ for soft-rock sites, $S_{hr}=1$ for hard-rock sites and $S_{sr}= S_{hr}= 0$ for alluvial sites.

Joyner and Boore (1981) developed a relation for peak ground velocity. This empirical attenuation relation was obtained by a regression analysis based on 38 data points from the 1979 Imperial

Valley earthquake and 68 data points from other earthquakes. That is:

$$\log V_m = -0.67 + 0.489M_w - \log R_s - 0.00256R_s + 0.17S + 0.22P \tag{3.2}$$

where $R_s = \sqrt{d^2 + 16}$, V_m is the peak ground velocity in cm/sec, d is the closest distance to the surface projection of the fault rupture (in km), S=0 and 1 for rock and soil sites respectively, and P=0 and 1 for the 50th percentile and the 84th percentile, respectively.

More recently, Kamiyama et al. (1992) developed semi-empirical relations for peak ground responses, which account for effects of local site condition. They found that the ground response depends on the earthquake magnitude, the hypocentral distance (R_s) and the amplification factor of the site (AMP(V)). For example, the peak ground velocity, V_m, can be estimated by:

$$V_m = \begin{cases} 2.879 \times 10^{0.153M} \times AMP(V) & R_s \leq 10^{0.014+0.218M} \\ 3.036 \times 10^{0.511M} \times AMP(V) / R_s^{1.64} & R_s > 10^{0.014+0.218M} \end{cases} \tag{3.3}$$

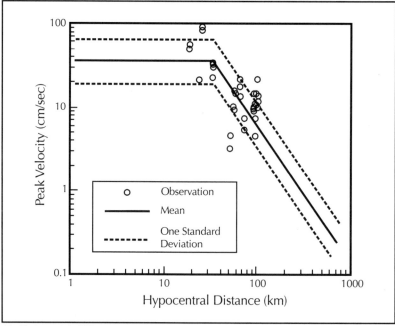

After Kamiyama et al., 1992

■ Figure 3.2 Comparison of Peak Ground Velocity at Rock Sites in the 1989 Loma Prieta Event with Kamiyama et al. Relation

For rock sites (AMP(V)=1), Figure 3.2 presents the semi-empirical relation by Kamiyama et al. and observed data from the 1989 Loma Prieta earthquake. Note that the amplification factors are available from response spectra of local sites, and vary with frequency content.

The most recent empirical relations for peak ground displacement were developed by Gregor (1995). He considers the effects of different fault mechanisms and wave types. For example, the peak rock displacement for shear waves (i.e., SH waves) due to strike-slip fault can be estimated by:

$$\log D_m = -5.0 + 1.02 M_w - 0.87 \log R_s \qquad (3.4)$$

Figure 3.3 shows data points for the 1989 Loma Prieta earthquake (M_w=7.0) and the mean attenuation curve (solid line) from Equation 3.4. The dashed lines are the mean curve ± one standard deviation.

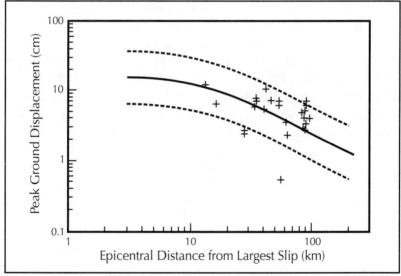

After Gregor, 1995

■ Figure 3.3 Comparison of Peak Ground Displacement at Rock Sites in the 1989 Loma Prieta Event with Gregor Relation

As shown in Figures 3.2 and 3.3, even with the "best" attenuation relation, observed values are only within a factor of two of predicted values.

E F F E C T I V E   P R O P A G A T I O N   V E L O C I T Y

Since pipelines are typically buried horizontally 1 to 3 m below the ground surface, both body and surface waves are of interest. The following sections focus on the techniques for estimating effective propagation velocity for body and surface waves.

3.3.1   B O D Y   W A V E S

For body waves, we consider herein only S-waves since S-waves carry more energy and tend to generate larger ground motion than P-waves. For the S-wave, the horizontal propagation velocity, that is the propagation velocity with respect to the ground surface, is the key parameter. For vertically incident S-waves, the apparent propagation velocity is infinity. However there is typically a small angle of incidence in the vertical plane leading to non-zero horizontal ground strain. M. O'Rourke et al. (1982) have studied the apparent horizontal propagation velocity, C, for body waves. They developed an analytical technique, utilizing all three components of motion at the ground surface and a ground motion intensity tensor, for evaluating the angle of incidence of S-waves. The apparent propagation velocity for S-waves is then given by:

$$C = \frac{C_s}{\sin \gamma_s} \tag{3.5}$$

where γ_s is the incidence angle of S-waves with respect to the vertical and C_s is the shear wave velocity of the surface soils.

Table 3.1 shows results by the ground motion intensity method for the 1971 San Fernando and the 1979 Imperial Valley events as well as values for other events from more direct techniques. Note that the apparent propagation velocity for S-waves ranged from 2.1 to 5.3 km/sec with a average of about 3.4 km/sec.

■ Table 3.1 Summary of Apparent Horizontal Propagation Velocities for S-waves

Event	Site Conditions	Focal Depth (km)	Epicentral Distance (km)	C (km/s)	Method for Calculating C
Japan 1/23/68	60 m soft Alluvium	80	54	2.9	Cross-correlation array with common time
Japan 7/1/68	60 m soft Alluvium	50	30	2.6	Cross-correlation array with common time
Japan 5/9/74	70 m of silty clay, sand & silty sand	10	140	5.3	Cross-correlation array with common time
Japan 7/8/74	70 m of silty clay, sand & silty sand	40	161	2.6	Cross-correlation array with common time
Japan 8/4/74	70 m of silty clay, sand & silty sand	50	54	4.4	Cross-correlation array with common time
San Fernando 2/9/71	Variable	13	29 to 44	2.1	Ground motion intensity tensor (median value)
Imperial Valley 10/15/79	>300 m Alluvium	Shallow	6 to 57	3.8	Ground motion intensity tensor (median value)
Imperial Valley 10/15/79	>300 m Alluvium	Shallow	6 to 93	3.7	Epicentral distance vs. initial S-wave travel time

3.3.2 SURFACE WAVES

For surface waves, we only consider R-waves since L-waves generate bending strains in buried pipelines which, particularly for moderate pipe diameters, are significantly less than axial strain induced by R-waves. As indicated previously, R-waves cause the ground particles to move in a retrograde ellipse within a vertical plane. The horizontal component of the ground motions for R-waves is parallel to the propagation path and thus will generate axial strain in a pipe laying parallel to the direction of wave propagation. Since R-waves always travel parallel to the ground surface, the phase velocity of the R-waves, C_{ph}, is the apparent propagation velocity.

Note that the phase velocity is defined as the velocity at which a transient vertical disturbance at a given frequency, originating at the ground surface, propagates across the surface of the medium. The phase velocity is a function of the variation of the shear wave velocity with depth, and, unlike body waves, is also a function of frequency. For R-waves, the wavelength λ, frequency f and the phase velocity C_{ph} are interrelated by:

$$C_{ph} = \lambda f \qquad (3.6)$$

The variation with frequency is typically quantified by a dispersion curve. Analytical and numerical solutions are available in the technical literature to generate dispersion curves for layered soil profiles (Haskell, 1953, Schwab and Knopff, 1977).

M. O'Rourke et al. (1984) developed a simple procedure for determining the dispersion curve for layered soil profiles in which the shear wave velocity increases with depth. Figure 3.4 presents a normalized dispersion curve for a uniform layer of thickness, H_s, with shear velocity C_L and Poisson's ratio, ν_L, over a half space with shear velocity C_H and Poisson's ratio, ν_H. The curves are for two values of the shear velocity ratio. The dispersion relationship is not strongly affected by the densities of the layer and half space and those parameters are excluded from Figure 3.4.

Considering first the simplest case of a uniform layer over a half space, they found that at low frequencies $\left(H_s f / C_L \leq 0.25\right)$, the wavelength is large compared to the layer thickness, and the phase velocity is slightly less than the shear wave velocity of the stiffer half space. That is, the R-wave is not greatly affected by the "thin" layer. Conversely, at high frequencies $(H_s f/C_L > 0.5)$, the wavelength is comparable to or smaller than the layer thickness, and the phase velocity is sightly less than the shear wave velocity of the layer. The dispersion curve for an arbitrary single layer over a half space can be approximated by:

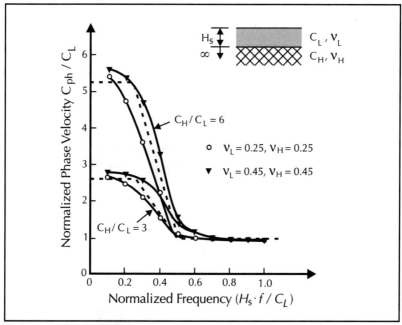

After M. O'Rourke et al., 1984

■ **Figure 3.4 Normalized Dispersion Curve for Single Layer Over Half Space**

$$C_{ph} = \begin{cases} 0.875\,C_H, & \dfrac{H_s f}{C_L} \le 0.25 \\[2ex] 0.875\,C_H - \dfrac{0.875\,C_H - C_L}{0.25}\left(\dfrac{H_s f}{C_L} - 0.25\right), & 0.25 \le \dfrac{H_s f}{C_L} \le 0.50 \\[2ex] C_L, & \dfrac{H_s f}{C_L} \ge 0.50 \end{cases}$$

$$(3.7)$$

where f is the frequency in Hz.

This technique can also be extended to multiple soil layers. For the two soil layer case shown in Figure 3.5(a), separate single layer models in Figure 3.5(b) for short wavelengths and in Figure 3.5(c) for long wavelengths are considered. The dispersion curves for each single layer model are combined to obtain the curve for the whole profile as shown in Figure 3.6.

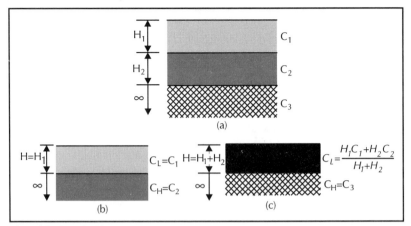

After M. O'Rourke et al., 1984

■ **Figure 3.5 Idealization of Complex Soil Profile**

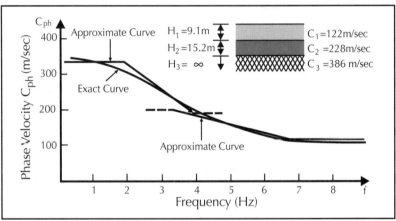

After M. O'Rourke et al., 1984

■ **Figure 3.6 Dispersion Curve for Two Layer Soil Profile**

WAVELENGTH

For typical soil profiles, in which the material stiffness increases with depth, the propagation or phase velocity of the fundamental R-waves is an increasing function of the wavelength. That is, long wavelength waves travel faster than short wavelength waves. Hence, the effective propagation velocity for strain calculations with stations separated by a distance, L_s, could very well be related to the phase velocity of R-waves having a wavelength λ proportional to L_s.

A precise analytical relationship between separation distance, L_s and the appropriate wavelength, λ, is complicated by the fact that the displacement amplitudes associated with various wavelengths are not constant. However, a reasonable starting point might be $2L_s<\lambda<4L_s$. As pointed out by Wright and Takada (1978), ground motion of two points due to a wave with $\lambda=2L_s$ would be out-of-phase by 180°, leading to fairly large relative displacements and strains. Similarly, ground motion at two points due to a wave with $\lambda=4L_s$ would be out-of-phase by 90°, again leading to relative displacements and strains. If $\lambda=L_s$, the ground motions would be in-phase, and there would be no contribution to relative displacements and strains due to that wavelength. Thus, the effective propagation velocity, C_{eff}, would appear to be the R-wave phase velocity, C_{ph}, for a wavelength equal to about 2-4 separation distances.

Figure 3.7 presents back calculated values for the effective propagation velocity during the 1971 San Fernando event for a number of stations at the northern end of the Los Angeles Basin, as well as the phase velocity for the fundamental R-waves, calculated for $\lambda=2L_s$ and $\lambda=4L_s$. The R-wave model with $\lambda=4L_s$ seems to provide a better overall match to the observed effective propagation velocity data than the R-wave model with $\lambda=2L_s$.

As shown in Figure 3.7, for separation distances less than about 500 m (1640 ft), the R-wave model with $\lambda=4L_s$ matches fairly well the observed effective propagation velocity data. For separation distances greater than 500 m the R-wave model with $\lambda=4L_s$ is conservative, i.e., an underestimation of the effective propagation velocity. However, these large separation distances are fast approaching the upper limit for engineering applications.

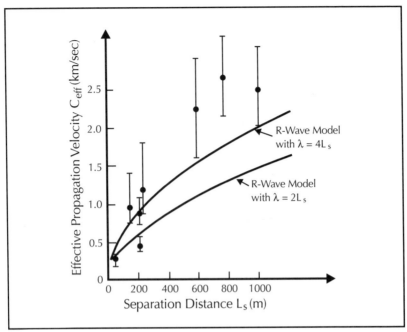

After M. O'Rourke et al., 1984

■ Figure 3.7 Effective Propagation Velocity vs. Separation Distance

G R O U N D   S T R A I N   A N D
C U R V A T U R E   D U E   T O
W A V E   P R O P A G A T I O N

For the analysis and design of buried pipelines, the effects of seismic wave propagation are typically characterized by the induced ground strain and curvature. Newmark (1967) developed a simplified procedure to estimate the ground strain. He considers a simple traveling wave with a constant wave shape. That is, on an absolute time scale, the acceleration, velocity and displacement time histories of two points along the propagation path are assumed to differ only by a time lag, which is a function of the separation distance between the two points and the speed of the seismic wave. For such a case, he shows that the maximum ground strain ε_g (tension and compression) in the direction of wave propagation is given by:

$$\varepsilon_g = \frac{V_m}{C} \qquad (3.8)$$

where V_m is the maximum horizontal ground velocity in the direction of wave propagation and C is the propagation velocity of the seismic wave.

Similarly, the maximum ground curvature, κ_g, that is the second derivative of the transverse displacement with respect to distance, is given by:

$$\kappa_g = \frac{A_m}{C^2} \qquad (3.9)$$

where A_m is the maximum ground acceleration perpendicular to the direction of wave propagation.

These two relations for ground strain and curvature along the direction of wave propagation are relatively straight forward. The ground motion parameters V_m and A_m, the maximum particle velocity and acceleration can be obtained from earthquake records or from attenuation relations, discussed previously. For R-wave propagation, the ground strain parallel to the ground surface is given by Equation 3.8 where C, as shown above, can be taken as the phase velocity corresponding to a wavelength equal to four times the separation distance. However, these relations for ground strain and curvature need to be modified if the direction of interest is not parallel to the direction of wave propagation.

Consider the case of S-waves. If the pipeline is orientated parallel to the direction of propagation, S-waves would induce bending in the pipeline. The corresponding ground curvature is given by Equation 3.9 where C is the apparent propagation velocity with respect to the ground surface given in Equation 3.5 and Table 3.1. If there is an angle in the horizontal plane between the pipe axis and the direction of propagation, there is a component of ground motion parallel to the pipe axis. The resulting ground strain along the pipe axis is a function of this angle in the horizontal plane. Yeh (1974) has shown that the ground strain is a maximum for an angle of 45° in the horizontal plane where

$$\varepsilon_g = \frac{V_m}{2C} \tag{3.10}$$

where C is the apparent propagation velocity with respect to the ground surface given in Equation 3.5 and Table 3.1.

For R-waves propagation, the horizontal ground strain along the direction of propagation is given by Equation 3.8 where C is the phase velocity for a wavelength equal to roughly four times the separation distance.

E F F E C T S   O F   V A R I A B L E   S U B S U R F A C E   C O N D I T I O N S

The ground strains and curvatures described previously are due to wave propagation effects. The apparent propagation velocity relations presented apply to relatively uniform soil layering in the horizontal direction. However as noted by Kachadoorian (1976) and Wang and M. O'Rourke (1978), damage to buried pipelines is often concentrated in areas with variable subsurface condition (i.e., non-uniform soil properties in a horizontal direction). In a more recent example, Hall (1995) notes relative large amount of buried pipeline damage during the 1994 Northridge event in areas where inclined ground surface or inclined soil-rock interface exists. It is believed that ground strain for sites with variable subsurface condition is due, in large part, to local differences in site response or site amplification.

An advanced finite element model for ground response at a valley was recently proposed by Zhang et al. (1995). They analyze ground response by using a hybrid numerical technique, which combines the Boundary Integral Equation with the Finite Element Method. The finite elements are for the complex site of interest, while the boundary integral equation considers the effects of infinite media beyond the complex site.

For given ground motions at rock sites, the particle motions, the phase velocity of surface wave and the ground strain can be obtained by this hybrid analysis. Using the record of the 1989 Loma Prieta earthquake and neglecting the liquefaction of sand fills, Zhang and Papageorgiou (1996) analyzed the ground response for the Marina District during the 1989 Loma Prieta earthquake. They found that the maximum ground acceleration and velocity may have reached values as high as 0.23 g and 34 cm/sec respectively, and that ground strains induced by wave propagation were of the order of 10^{-4}.

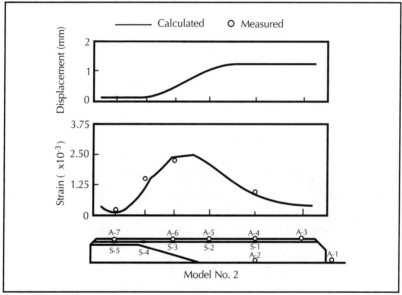

After Nishio et al., 1983

■ Figure 3.8 Axial Strains and Model for a Half Valley

3.6.1 NUMERICAL MODELS

Nishio et al. (1983) carried out a series of laboratory tests to study the amplification response of ground due to the inclined soil-rock interface. Figures 3.8 and 3.9 show, for example, two basic models they considered. The bottom of the model was shaken as a unit, corresponding to vertically incident waves (i.e., no horizontal wave propagation effects). For the single inclined subsurface

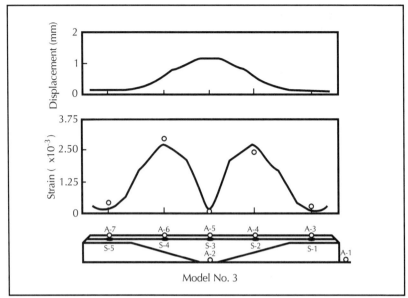

After Nishio et al., 1983

■ Figure 3.9 Axial Strains and Model for a Valley

in Model No. 2 in Figure 3.8, the ground strain was largest near the inclined surface. For the valley situation (i.e., two inclined sub-surfaces) in Model No. 3 in Figure 3.9, the ground strain was roughly symmetric about center of the valley and ground strain was also largest over the inclined subsurface. They also performed a finite element analysis for these models. As shown in Figures 3.8 and 3.9, the numerical results match well with the experimental results for East-West Component at the Shiogama site recorded during the 1978 Off Miyagi Ken Oki earthquake.

For estimating the ground response of complex sites, ap-proaches using 1-D, 2-D or 3-D finite element techniques have been pursued. For example, Ando et al. (1992) use a 2-D finite element program to analyze the dynamic response of the site shown in Figure 3.10. In this figure, the shaded area is embanked ground (fill deposits) with a shear velocity of 166 m/s while the shear ve-locity for the original ground is 300 m/s.

Using the 1978 Off Miyagi Prefecture earthquake record which had a maximum ground acceleration of 25 gal, Ando et al. deter-mined the ground and pipe strains as shown in Figure 3.11.

Again, all points along the base of the model had the same motion without a time lag, modeling vertically propagating body

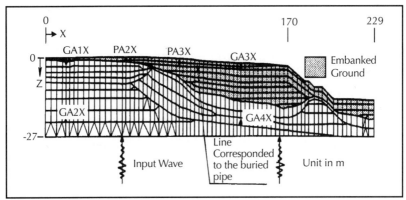

After Ando et al., 1992

■ **Figure 3.10 Profile of a Complex Site**

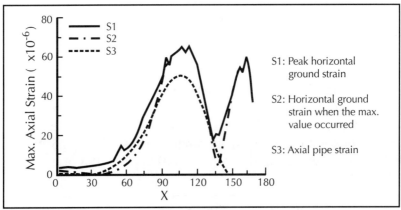

After Ando et al., 1992

■ **Figure 3.11 Distribution of Strains For Complex Site**

waves. The ground strain was largest where the original ground surface is inclined (i.e., near PA3X), and where the fill deposits were inclined (i.e., to the right of location mark 170 m). The maximum ground strain due to the complex site is about 10 times that for original uniform ground. The numerical results match well with the observed values during the same earthquake.

In order to test their simplified approach which will be discussed later, Liu and M. O'Rourke (1997a) developed a numerical approach for calculating the ground response at a site similar to Nishio et al.'s Model 2 shown in Figure 3.8. Different from the Nishio et al. model, Liu and M. O'Rourke considered the effects of material outside the immediate area of interest, as shown in

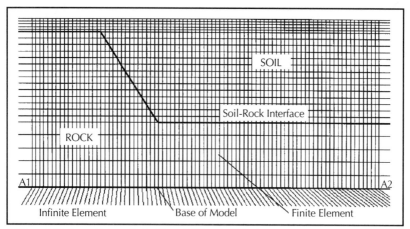

■ Figure 3.12 Area of Interest in the Liu and O'Rourke Finite Element Model

Figure 3.12. Specifically, infinite elements are used in order to eliminate the reflection at the outside boundary. Shear waves are generated by inputting acceleration records or prescribed displacements along A1-A2 line, and propagating vertically toward the ground surface. Above the darkened line is a soil layer with shear velocity ranged from 150 m/s to 1000 m/s, Poisson ratio of 0.32 and soil density of 2000 kg/m^3 (120 pcf), while below the darken line is rock with shear velocity ranging from 1250 to 2500 m/s, Poisson ratio of 0.20 and density of 3200 kg/m^3 (192 pcf).

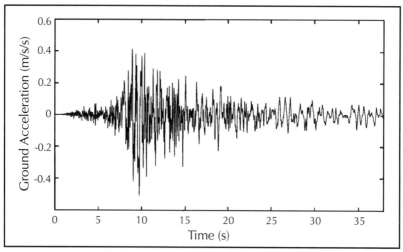

■ Figure 3.13 Horizontal Component of Acceleration Record

Three acceleration records are inputted at the base of the model. They are the one shown in Figure 3.13, and two other records at Stations 102 and 211 from the 1971 San Fernando earthquakes. The relatively high frequency motion is considered to be representation of rock motion at depth. The time increment for the digitized acceleration input is 0.01 s.

Liu and M. O'Rourke analyze the peak ground response for a soil-rock interface angle $\alpha=10°$, $\alpha=45°$, and $\alpha=90°$ as shown, for example, in Figure 3.14.

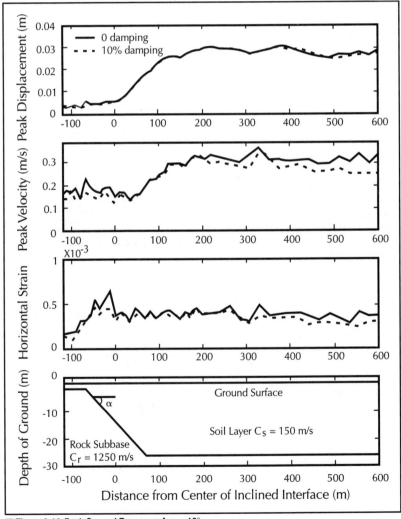

■ Figure 3.14 Peak Ground Response for $\alpha=10°$

Figures 3.14, 3.15 and 3.16 present peak ground displacement, peak ground velocity and peak horizontal ground strain near the ground surface for three inclined rock models with inclination angle, α, of 10°, 45° and 90°, respectively. The shear wave velocity is 150 m/s for the surface soil layer and 1250 m/s for the rock in those cases. As shown in these three figures, the ground displacement and velocity are small over the shallow soil layer and large over the deeper soil layer, that is generally increasing with increas-

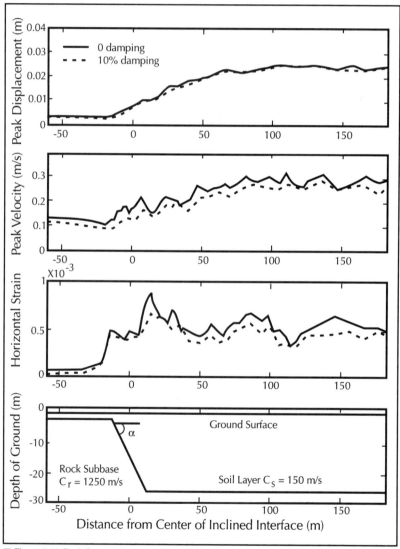

■ Figure 3.15 Peak Ground Response for α=45°

ing thickness of the soil layer. The ground strain is typically small for the soil layer to the left of inclined soil-rock interface, and reaches a maximum value around the inclined soil-rock interface. This maximum ground strain then decreases with the increase of the distance away from the inclined soil-rock interface. This ground strain is due, in part, to the waves deflected from the inclined rock-soil subsurface and reflecting between the ground surface and the horizontal rock-soil subsurface.

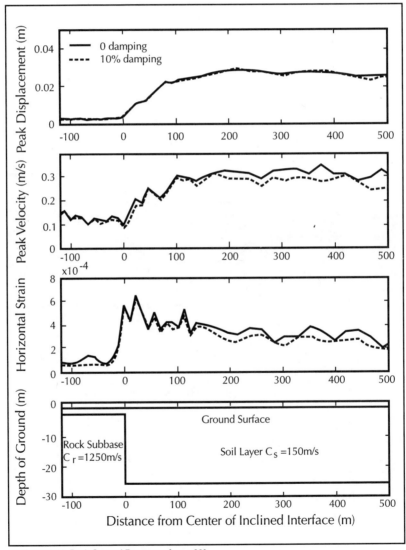

■ Figure 3.16 Peak Ground Response for $\alpha=90°$

3.6.2 SIMPLIFIED MODEL

Liu and M. O'Rourke (1997a) have proposed a simplified model for estimating the ground strain for a site with inclined soil-rock interface as shown in Figure 3.17, in which shear waves are propagating vertically from below.

In Figure 3.17, D_A is the peak ground displacement at the shallow soil layer while D_B is the corresponding value for the deep soil layer. A simple estimate for ground strain over the inclined rock surface is the difference in the ground displacement divided by the separation distance. Assuming that the soil motions are in phase and taking the horizontal projection of the inclined rock surface, L_{AB}, as the separation distance.

$$\varepsilon_g = \frac{D_B - D_A}{L_{AB}} \tag{3.11}$$

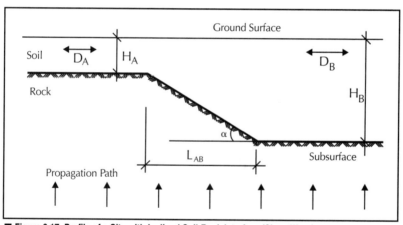

■ Figure 3.17 Profile of a Site with Inclined Soil-Rock Interface (Shear Wave)

A comparison of results from this simple relation with the numerical results given, for example, in Figures 3.14 and 3.15 indicated that Equation 3.11 provides reasonable estimates as long as the inclination angle α is less than 45°. For inclination angles greater than about 45°, Equation 3.11 overestimates peak ground strains from the numerical model. Note in this regard that for a vertical rock face (i.e., α=90°), L_{AB} is zero and Equation 3.11 gives infinite ground strain which, of course, is not realistic.

In addition, the numerical results show that with the increase in the separate distance L_{AB}, the ground strain decreases to a certain value and then remains constant. That is, for very large L_{AB} (i.e., very small inclined angle), the ground strain is a non-zero constant as opposed to a decreasing function of L_{AB} as predicted by Equation 3.11.

Based upon these comparisons, Equation 3.12 is suggested for estimate of ground strain over an inclined rock surface.

$$\varepsilon_g = \begin{cases} \left(D_B - D_A\right) \cdot \left(\dfrac{0.5 \cdot \tan \alpha}{H_B - H_A} + \dfrac{\pi}{2TC_s}\right) & \alpha \leq 45° \\[4mm] \left(D_B - D_A\right) \cdot \left(\dfrac{0.5}{H_B - H_A} + \dfrac{\pi}{2TC_s}\right) & \alpha > 45° \end{cases}$$

(3.12)

where H_A is the thickness of the shallow soil layer, H_B is the thickness of the deeper soil layer and T is the predominant period at the ground surface over the deep soil layer.

For an inclination angle $\alpha \leq 45°$, the peak ground strain is an increasing function of the inclined angle. For $\alpha > 45°$, ground strain is taken as the value for a 45° inclination angle.

3.6.3 COMPARISON

In order to test this simplified approach, Liu and M. O'Rourke calculated the ground strains for different sites with $\alpha \geq 3°$, $H_A \leq 5.0\,\text{m}$ (16 ft), $H_B \geq 10\,\text{m}$ (33 ft), 150 m/s $\leq C_{s-soil} \leq 1000$ m/s, $C_{s-rock} \geq 1000$ m/s, by using the finite element model in Figure 3.12. For example, Figure 3.18 shows the peak ground strain as a function of length L_{AB} for $C_{s-soil}=150$ m/s, $H_A=2.0$ m (6.6 ft), $H_B=26$ m (85 ft) and $C_{s-rock}=1250$ m/s, while Figure 3.19 shows results for the same model with $C_s=210$ m/s.

Another type of comparison is shown in Figure 3.20. There the numerical strain from the model in Figure 3.12 with $\alpha \geq 3°$ is plotted versus the estimated strain from Equation 3.12. The ground strains for the same model subject to the 1971 San Fernando earthquake are also included in this figure. Note that for the purposes of comparison, the two San Fernando time histories were scaled so that the peak accelerations matches that in Figure 3.13.

Also shown in Figure 3.20 are the numerical results (2.5×10^{-3}) for the model in Figure 3.8 (Nishio et al. 1983) as well as the corresponding value from Equation 3.12. For the model considered ($C_s=50$ m/s, $T=0.016$ s), the estimated strain from Equation 3.12 is 2.7×10^{-3}.

The ground strains from the recent numerical approach (dot points) are generally less than that from the simplified approach except for small strain cases. Hence, Equation 3.12 can be used to estimate ground strain at a site with an inclined soil-rock interface.

As shown in Figures 3.18 through 3.20, the ground strain from Equation 3.12 is somewhat less than that from the numerical approach when the separation distance is large. Overall, however, the match is reasonably good and the estimated strain is conservative at large strains.

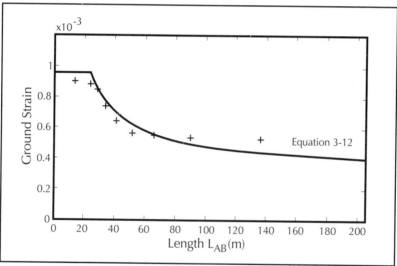

■ Figure 3.18 Comparison of Ground Strain for $C_s=150$ m/s

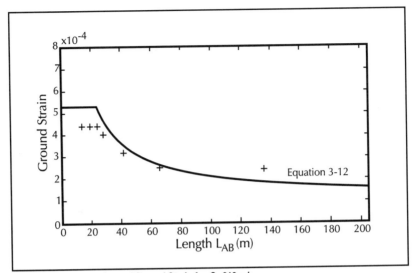

■ Figure 3.19 Comparison of Ground Strain for C_s=210 m/s

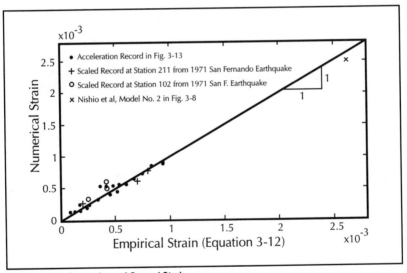

■ Figure 3.20 Comparison of Ground Strain

PIPE FAILURE MODES AND FAILURE CRITERION

This chapter describes the failure modes for buried pipelines subject to seismic loading. The principal failure modes for corrosion-free continuous pipelines (e.g. steel pipe with welded joints) are rupture due to axial tension, local buckling due to axial compression and flexural failure. If the burial depth is shallow, continuous pipelines in compression can also exhibit beam-buckling behavior. Failure modes for corrosion-free segmented pipelines with bell and spigot type joints are axial pull-out at joints, crushing at the joints and round flexural cracks in pipe segments away from the joints. For each of these failure modes, the corresponding failure criterion is presented, first for continuous pipelines and then for segmented pipelines.

4.1 CONTINUOUS PIPELINE

The principal failure modes for corrosion-free continuous pipeline with burial depth of about three feet or more are tensile rupture and local buckling. Buried pipelines with burial depths less than about 3 feet (i.e., shallow trench installation) may experience beam buckling behavior. Beam buckling has also occurred during post earthquake excavation undertaken to relieve compressive pipe strain.

4.1.1 TENSILE FAILURE

When strained in tension, corrosion free steel pipe with arc welded butt joints is very ductile and capable of mobilizing large strains associated with significant tensile yielding before rupture.

On the other hand, older steel pipe with gas-welded joints often cannot accommodate large tensile strain before rupture. In addition, as discussed in detail in Section 4.1.4, welded slip joints in steel pipe do not perform as well as butt welded joints. The 1994 Northridge event provides a case history of these differences in behavior. According to T. O'Rourke and M. O'Rourke (1995), none of the four arc-welded steel pipes with butt joints along Balboa Blvd. suffered tensile rupture when subjected to longitudinal PGD. However three gas-welded pipes with slip joints suffered tensile rupture when subjected to the same PGD.

The strain associated with tensile rupture is generally well above about 4% (Newmark and Hall, 1975). Often an ultimate tensile value of 4% is used, beyond which the pipeline is considered to have failed in tension.

Analytical methods for post-yield performance require full descriptions of the stress-strain behavior. One of the most widely used models is the one proposed by Ramberg and Osgood (1943). The Ramberg Osgood model is given by:

$$\varepsilon = \frac{\sigma}{E}\left[1 + \frac{n}{1+r}\left(\frac{\sigma}{\sigma_y}\right)^r\right]$$

(4.1)

where ε is the engineering strain, σ is the uniaxial tensile stress, E is the initial Young's modulus, σ_y is the apparent yield stress, n and r are Ramberg Osgood parameters. Commonly used values for σ_y, n and r for various grades of steel are listed in Table 4.1. The Ramberg Osgood relationship will be used in determining the response of continuous pipe subject to longitudinal PGD in Chapter 6.

■ Table 4.1 Ramberg Osgood Parameters for Mild Steel and X-grade Steel

	Grade-B	X-42	X-52	X-60	X-70
Yield Stress (MPa)	227	310	358	413	517
n	10	15	9	10	5.5
r	100	32	10	12	16.6

4.1.2 LOCAL BUCKLING

Buckling refers to a state of structural instability in which an element loaded in compression experiences a sudden change from a stable to an unstable condition. Local buckling (wrinkling) involves local instability of the pipe wall. After the initiation of local shell wrinkling, all further geometric distortion caused by ground deformation or wave propagation tends to concentrate at the wrinkle. The resulting large curvatures in the pipe wall often then lead to circumferential cracking of the pipe wall and leakage. This is a common failure mode for steel pipe. Wave propagation in the 1985 Michoacan event caused this type of damage to a water pipe in Mexico City. Permanent ground deformation caused this type of damage to a liquid fuel pipeline in the 1991 Costa Rica event, as shown in Figure 4.1, and to water and gas pipelines in the 1994 Northridge event.

After M. O'Rourke and Ballantyne, 1992

■ **Figure 4.1 Local Buckling to RECOPE Pipeline**

Based on prior laboratory tests on thin wall cylinders, Hall and Newmark (1977) suggest that compressional wrinkling in a pipe normally begins at a strain of 1/3 to 1/4 of the theoretical value of:

$$\varepsilon_{theory} = 0.6 \cdot t/R \qquad (4.2)$$

where t is the pipe wall thickness and R is the pipe radius. Hence, in terms of failure criterion the onset of wrinkling occurs at strains in the range:

$$0.15 \cdot t / R \leq \varepsilon_{cr} \leq 0.20 \cdot t / R \qquad (4.3)$$

This assumed wrinkling strain is thought to be appropriate for thin wall pipe but somewhat conservative for thicker wall pipe. The additional amount of longitudinal compressive deformation across the wrinkled zone which results in tearing of the pipe wall due to large curvature at individual wrinkles is, at present, not well established.

4.1.3 BEAM BUCKLING

Beam buckling of a pipeline is similar to Euler buckling of a slender column in which the pipe/column undergoes a transverse upward displacement. The relative movement is distributed over a large distance and hence the compressive pipe strains are not large. As a result beam buckling of a pipeline in a ground compression zone is considered more desirable than local buckling since the potential for tearing of the pipe wall is lessened.

Beam buckling of pipes has been observed in a few events. For example, during the period from 1932 to 1959, displacements on the order of 360 mm (14 in.) accumulated across the Buena Vista reverse fault (Howard, 1968). This ground movement led to compression stresses in oil pipelines, ranging from 51 to 406 mm (2 to 16 in.) in diameter. The oil pipelines buried at depths between 0.15 and 0.30 m (6 to 12 in.), and in loose to medium soil, lifted out of the ground as a result of compressive forces.

Another example occurred during the 1979 Imperial Valley earthquake. Two high pressure pipelines, 219 mm (8.6 in.) and 273 mm (10.7 in.) in diameter, crossing the main trace of the Imperial fault were affected. No evidence of local buckling or beam buckling was observed immediately after the event. However, removal of cover during inspection after the earthquake caused both pipes to displace laterally in a beam buckling mode (McNorgan, 1989).

As opposed to tensile rupture, or wrinkling and associated tearing of the pipe wall, the pipes do not "fail" after beam buckling. The beam bucking of pipes may better be described as a serviceability problem since the pipe continues to serve its function of transmitting fluid without interruption. In that sense it is difficult to establish a failure criterion for beam buckling strictly in terms of pipe material properties. Its occurrence depends on several factors such as the bending stiffness and burial depth of the pipe as well as initial imperfections. A discussion of the conditions leading to beam buckling is presented below.

Beam buckling of buried pipelines has been the subject of many analytical studies. An analytical solution for a type of beam buckling was first provided by Marek and Daniels (1971). They studied the behavior of continuous crane rails subject to a temperature rise. Hobbs (1981) adapted the Marek and Daniels model to solve the problem of buckling of submarine pipelines. Hobbs considered the compressive loads due to temperature changes or internal pressures, which can cause beam buckling in the presence of an initial imperfection. His model is shown in Figure 4.2. where w is the submerged weight of pipeline per unit length, L_b is the length of the buckle, L_s is the length of pipe adjacent to the buckle which slips with respect to the surrounding soil.

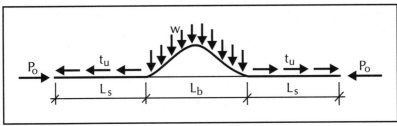

After Hobbs, 1981

■ **Figure 4.2 Vertical Buckling Mode**

Figure 4.3 shows the buckling load versus the buckling length, L_b, and the maximum buckling amplitude, y_o.

As shown in Figure 4.3, the buckling load is a nonlinear function of buckling length and achieves a minimum value at Point B for a certain buckling length, L_{bm}. Due to uplift resistance and soil

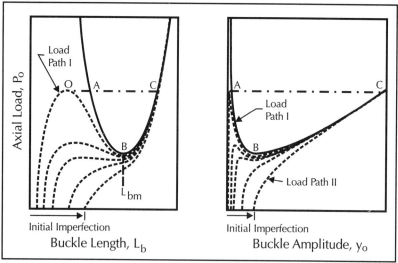

After Hobbs, 1981

■ **Figure 4.3 Vertical Buckling Load and Equilibrium Paths for Increasing Imperfection Level**

friction, the buckling load increases for the buckling length larger than L_{bm}. Obviously, for any load higher than that at Point B, there are two possible buckling lengths and amplitudes, for example, values at Point A and Point C in Figure 4.3. The situation at Point A is unstable and the pipe will tend to eventually snap through to Point C at a constant axial load. Note that, for the case of small initial imperfection, the pipe buckles and the post-buckling curve is shown, for example, in Path I after the compressive load reaches the maximum value at Point O. On the contrary, for large initial imperfection case, no snap-through is observed (for example, Path II) since the imperfection is gradually magnified in this case.

Kyriakides et al. (1983) used a somewhat similar approach more recently which requires imperfection information. Ariman and Lee (1989) augmented the Kyriakides et al.'s model by considering an axial friction force at the pipe-soil interface, nonlinear uplift resistance, as well as an elasto-plastic moment-curvature relation for the pipe. One difficulty in applying both the Kyriakides et al. or the Ariman and Lee model in practice, is that information on typical pipeline imperfections is apparently not readily available.

Meyersohn (1991) overcame this difficulty by extending Hobbs procedure to the problem of beam buckling of buried pipelines subject to longitudinal PGD. Figure 4.4 shows the distribution of axial compressive forces before buckling and after buckling. The length of the stressed portion of the pipe is directly associated with the magnitude and extent of the ground displacement. Once the axial force in the pipe P equals or exceeds a certain value, referred to as P_{max}, beam buckling occurs. The frictional forces are then relieved over the uplifted length of the pipe. The frictional forces also change direction over portions of the pipe at both sides of the buckle (i.e., the zone of reversed friction in Figure 4.4(b)), relieving some axial stress. The force P_o represents the maximum axial force after equilibrium is restored.

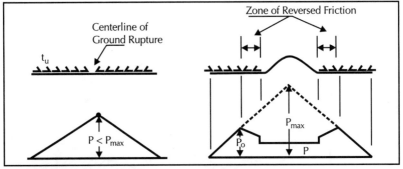

After Meyersohn, 1991

■ Figure 4.4 Distribution of Axial Compressive Forces

Intuitively, beam buckling is more likely to occur in pipelines buried in shallow trenches and/or backfilled with loose materials. That is, beam buckling load is an increasing function of the cover depth. Hence, if a pipe is buried at a sufficient depth, it will develop local buckling before beam buckling. Based on this concept, Meyersohn (1991) determined a critical cover depth by setting the lowest beam buckling stress equal to local buckling stress. Any pipe buried with less cover than the critical depth would experience beam buckling before local buckling. Conversely, if the pipe is buried at a depth more than the critical depth, it will experience local buckling. Figure 4.5 shows the critical cover depth for Grade B and X-60 steel pipes.

The shaded areas in the figure correspond to different degrees of backfill compaction. Note that critical depth for X-60 steel is larger than that for Grade B steel. That is, the stronger the pipe, the

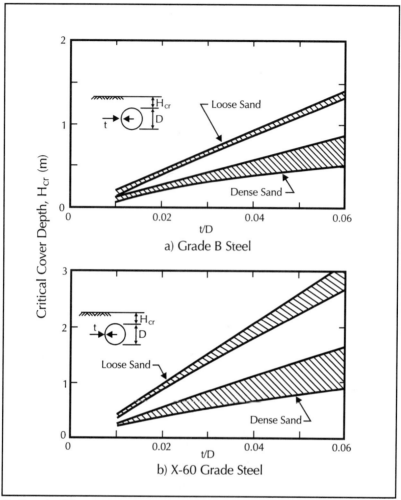

Critical Cover Depth, H_{cr} (m)

a) Grade B Steel

b) X-60 Grade Steel

After Meyersohn, 1991

■ **Figure 4.5 Analytical Critical Cover Depth of Pipe for Grade B and X-60 Steel**

smaller the possibility of shell wrinkling as opposed to beam buckling. However, as noted by Meyersohn (1991), the t/D ratio is
typically less than or about equal to 0.02. Hence, from Figure 4.5,
the likelihood of beam buckling of buried pipelines is small since
the critical depth is less than typical burial depths.

4.1.4 WELDED SLIP JOINTS

The failure criterion for steel pipelines with arc-welded butt joints is based on the strength of pipe material as discussed previously. However, for steel pipelines with slip joints, riveted joints or oxy-acetylene/gas welded joints, failure criterion is based on the strength of these joints since it is less than that of pipe materials. Many such steel pipelines have suffered failure at those joints during past earthquakes. For example, during the 1971 San Fernando earthquake, the Granada Trunk line (1260 mm in diameter) was damaged at its welded slip joints (T. O'Rourke and Tawfik, 1983). Figure 4.6 shows a slip joint where t is the pipe wall thickness.

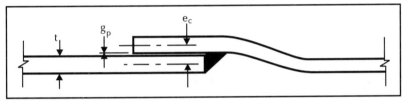

After Brockenbrough, 1990

■ **Figure 4.6 Slip Joint with Inner Weld**

Tawfik and T. O'Rourke (1985), Moncarz et al. (1987), and Brockenbrough (1990) analyzed the strength of slip joints. Considering 108 inch (2.74 m) diameter pipe with an inner weld, Moncarz et al. calculated a joint efficiency of 0.4 (strength of joint compared to strength of pipe) by using an inelastic finite element model. Considering the same type of joints, Brockenbrough determined a joint efficiency of 0.35, which is a little less (12%) than Moncarz et al.'s result. Figure 4.7 presents the joint efficiency of bell-spigot joints with an inner weld by Brockenbrough's model. Note that the maximum efficiency is 0.41, when no space or gap exists between the bell and spigot walls, and the joint efficiency is a decreasing function of the gap.

However, for most pipelines, slip joints with an outer weld are used since it is difficult to weld inside portions of pipe segments when the pipe diameter is small. Figure 4.8 shows a slip joint with an outer weld.

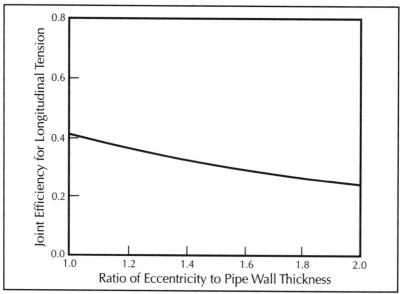

■ Figure 4.7 Joint Efficiency for Slip Joints with Inner Weld

As shown in Figure 4.8(b), t is the wall thickness, R_o is the pipe radius, e is the eccentricity (offset of the joint) and I is the length of curved portion of bell. Using an inelastic shell model, Tawfik and T. O'Rourke (1985) calculate joint efficiencies as shown in Figure 4.9. Note that two failure modes are considered in their analyses. Mode I refers to yielding in the vicinity of welded connections and Mode II refers to plastic flow in the curvilinear, belled

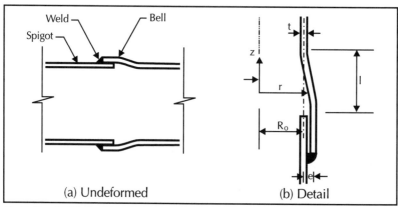

■ Figure 4.8 Slip Joint with Outer Weld

CHAPTER 4

ends of the joints. For normalized length of the bell $l/R_0 > 0.30$, Mode I governs and the joint efficiency is about 0.29.

As shown in Figure 4.7 and Figure 4.9, the joint efficiency of slip joints with an inner weld is often larger than that with an outer weld. Presumably this is because the eccentricity of the weld with respect to the pipe radius for joints with an inner weld are smaller than that with an outer weld.

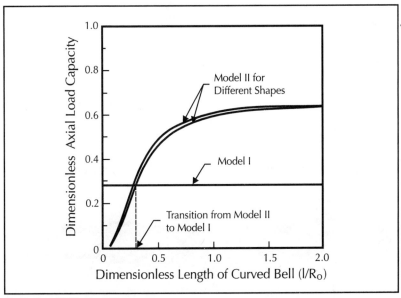

After Tawfik and T. O'Rourke, 1985

■ **Figure 4.9 Joint Efficiency of Slip Joints with an Outer Weld**

SEGMENTED PIPELINE

For segmented pipelines, particularly those with large diameters and relatively thick walls, observed seismic failure is most often due to distress at the pipe joints. For example, in the 1976 Tangshan earthquake, Sun and Shien (1983) observed that around 80 percent of pipe breaks were associated with joints. M. O'Rourke and Ballantyne (1992) identified six types of damage mechanism to segmented pipelines shown in Figure 4.10 during the 1991 Costa Rica earthquake. For the CI and DI transmission pipelines in the Limon area, 52% repairs are due to pull-out at joints (Figure 4.10(f)) and 42% repairs are due to breaks at segments (Figure 4.10(a)).

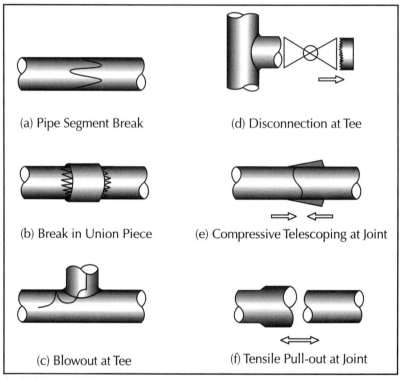

(a) Pipe Segment Break

(b) Break in Union Piece

(c) Blowout at Tee

(d) Disconnection at Tee

(e) Compressive Telescoping at Joint

(f) Tensile Pull-out at Joint

After M. O'Rourke and Ballantyne, 1992

■ **Figure 4.10 Damage Mechanisms for Segmented Pipelines**

Axial pull-out, sometimes in combination with relative angular rotation at joints, is a common failure mechanism in areas of tensile ground strain since the shear strength of joint caulking materials is much less than the tensile strength of the pipe. In areas of compressive ground strain, crushing of bell and spigot joints is a fairly common failure mechanism in, for example, concrete pipes. For small diameter segmented pipes, circumferential flexural failure have been observed in areas of ground curvature. For example, as observed by T. O'Rourke et al. (1991), more than 80 percent of the breaks in cast iron pipes with small diameters (100 mm to 200 mm (4 to 8 in)) in the Marina district after the 1989 Loma Prieta event were round cracks in pipe segments close to joints.

4.2.1 AXIAL PULL-OUT

In terms of failure criterion, information for the various types of segmented pipes is not as well developed as for continuous pipes. El Hmadi and M. O'Rourke (1989) summarized the then available information on joint pull-out failure. Specifically, based on laboratory tests by Prior (1935), El Hmadi and M. O'Rourke (1989) established a cumulative distribution for leakage as a function of the normalized joint axial displacement u_j^u/d_p shown in Figure 4.11. Note that u_j^u is the joint opening and d_p is the joint depth.

As shown in Figure 4.11, the mean value of the joint opening corresponding to joint leakage is $0.52\, d_p$ with a coefficient of variation of 10%. Hence, El Hmadi and M. O'Rourke suggest a relative joint displacement corresponding to 50% of the total joint depth as the failure criterion for pull-out of segmented pipelines with "rigid" joints.

For ductile iron pipes with rubber gasketed joints, they present laboratory data and semi-empirical relations developed by Singhal (1983) for the ultimate axial tensile force in the joint at pull-out.

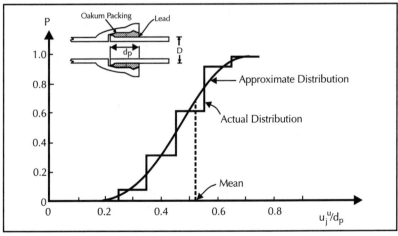

El Hmadi and M. O'Rourke, 1989

■ Figure 4.11 Cumulative Distribution Function for Leakage of Lead Caulked Joints

More recently, laboratory tests on concrete cylinder pipes with rubber gasketed joints by Bouabid and M. O'Rourke (1994) suggest that, at moderate internal pressures, the relative joint displacement leading to significant leakage corresponds to roughly half the total joint depth. Hence, it would appear that a relative axial joint extension of roughly half the total joint depth may be an appropriate failure criterion for many types of segmented pipes.

4.2.2 CRUSHING OF BELL AND SPIGOT JOINTS

As noted by Ayala and M. O'Rourke (1989), most of the concrete cylinder pipe failures in Mexico City occasioned by the 1985 Michoacan event were due to joint crushing. The corresponding failure criterion, based on laboratory tests for crushing of bell and spigot joints, is apparently not well established at this time.

According to Bouabid and M. O'Rourke's observation in their 1993 axial compressive tests, joint failure to reinforced concrete cylinder pipes with rubber gasketed joints can start at either the inner concrete lining or the outer concrete lining. That is, a circumferential crack starts to form in the ends of the concrete lining when the applied load nears the ultimate value. After concrete lining cracks, the critical section then becomes the welded interface between the steel joint ring and the steel pipe cylinder. The eccentricity existing between these two-elements causes some denting (or even local buckling) near this welded region. Such damaging action eventually would result in a leakage path and/or cause the section to burst. Hence, both Bouabid and M. O'Rourke (1994) as well as Krathy and Salvadori (1978) proposed that the crushing failure criterion for concrete pipes can be taken as the ultimate compression force of the concrete core at joints, F_{cr}. That is,

$$F_{cr} = \sigma_{comp} \cdot A_{core} \qquad (4.4)$$

where σ_{comp} is the compressive strength of concrete and A_{core} is the area of the concrete core. For plain concrete pipes, A_{core} is the cross-section area, while for reinforced pipes, the transformed area of steel bars needs to be added.

4.2.3 CIRCUMFERENTIAL FLEXURAL FAILURE AND JOINT ROTATION

When a segmented pipeline is subject to bending induced by lateral permanent ground movement or seismic shaking, the ground curvature is accommodated by some combination of rotation at the joints and flexure in the pipe segments. The relative contribution of these mechanisms depends on the joint rotation and pipe segment flexural stiffnesses. For a flexible pipeline system such as DI pipe with Tyton joints or FLEX joints, stress in the pipe segments starts to increase greatly only after the joint rotation capacity, typically about 4° and 15° respectively, is exceeded. On the other hand, for a more rigid segmented pipeline system such as cast iron pipe with cement/lead joints, ground curvature is accommodated from the start by some combination of joint rotation and flexure in the segments as will be discussed in more detail in Chapter 11.

In terms of failure criterion, it seems reasonable to base joint rotation failure/leakage criterion for "standard" segmented pipeline joints on some multiple (say 1.1 to 1.5) of the allowable angular offset for pipe laying purposes contained in manufacturer's literature. Table 4.2 contains a listing of such manufacturer's recommended allowable offset.

For cast iron or asbestos cement pipes subject to ground curvature, round flexural cracks in segments are a major failure mode. On the other hand, for concrete pipes subject to ground curvature, cracks typically occur at the bell and spigot ends due in part to the joint ring eccentricity mentioned previously.

For round flexural cracks, it seems reasonable to use, as a failure criterion, the pipe curvature corresponding to the smaller of the ultimate tensile or compressive strains for the material. In this regard El Hmadi and M. O'Rourke (1989) presented a listing of these mechanical properties for CI and DI pipe materials. Table 4.3 summarizes this information as well as the properties for other common pipe materials.

■ Table 4.2 Typical Manufacturer's Recommended Allowable Angular Offset (deg. and min.) for Various Pipe Joints

D (in)	Cast Iron	Ductile Iron Push-on	Mechanical	Prestressed Concrete	Concrete
4	4-00	5	8-18		
6	3-30	5	7-07		
8	3-14	4	5-21		
10		4	5-21		
12	3-00	4	5-21		
14		3	3-35		
16	2-41	3	3-35		2-19
18		3	3-00		2-04
20	2-09	3	3-00		1-52
24	1-47	2	2-23		1-34
27		2	2-23		1-24
30	1-26	2	2-23	1-44	1-15
33		1-30		1-35	1-09
36		1-30	2-05	1-28	1-03
42		1-30	2-00	1-16	1-03
48		1-30	2-00	1-06	1-03
60		1-30		0-56	
72		1-30		0-56	

■ Table 4.3 Mechanical Properties for Common Pipe Materials

Item	Cast Iron	Ductile Iron	Concrete	PVC	HDPE
Yield Strain $\times 10^{-3}$	1.0 to 3.0	1.75 to 2.17	0.1 to 1.3	17 to 22	22 to 25
Ultimate Strain $\times 10^{-3}$	5.0 to 40	100	0.25 to 3.0	50 to >100	50 to >100
Yield Stress (ksi)	14 to 42	42 to 52	0.32 to 4.0	5.0 to 6.5	2.2 to 2.5
Initial Modulus (ksi)	14,000	24,000	3,000	290-560	100-120

5

S O I L - P I P E
I N T E R A C T I O N

Buried pipelines are damaged in earthquakes due to forces and deformation imposed on them through interactions at the pipe-soil interface. That is, the ground moves and thereby causes the pipe to deform. For purposes of analysis, any arbitrary ground deformation can be decomposed into a longitudinal component (soil movement parallel to the pipe axis) and a transverse component (soil movement perpendicular to the pipe axis). Both those types of pipe-soil interactions are discussed in this chapter. In the transverse direction, interaction involves relative deformation and loading in both the horizontal and vertical planes. For relative ground movement in the vertical direction, one must distinguish between upward and downward pipe movement since the interaction forces are different for these two cases. Finally one must distinguish between pipe surrounded by competent, non-liquefied soil, and pipelines located in a liquefied layer.

5.1

C O M P E T E N T
N O N - L I Q U E F I E D   S O I L

Soil interaction forces for a pipeline surrounded by competent, non-liquefied soil are well established. They are based upon laboratory tests. For example, Trautmann and T. O'Rourke (1983) established a force-deformation relation for horizontal lateral movement as shown in Figure 5.1.

The ASCE Technical Council on Lifeline Earthquake Engineering (TCLEE) Committee on Gas and Liquid Fuel Lifelines (ASCE, 1984) have suggested, for the purpose of analysis, idealized elasto-plastic models as shown in Figure 5.2. Note that the elasto-plastic model is fully characterized by two parameters; the maximum re-

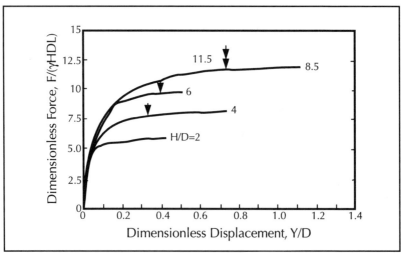

After Trautmann and T. O'Rourke, 1983

■ **Figure 5.1 Lateral Load-Deformation Relation at Pipe-Soil Interface**

sistance t_u, p_u, q_u, in the horizontal axial, horizontal transverse and vertical transverse directions respectively, having units of force per unit length and the maximum elastic deformation x_u, y_u, z_u, having units of distance. The equivalent elastic soil spring coefficients, having units of force per unit area, is simply the ratio of the maximum resistance divided by a half of the maximum elastic deformation, for example $2t_u/x_u$ for the horizontal axial (longitudinal) case. Note that this spring coefficient is effective only for relative displacements less than the maximum values of x_u, y_u, z_u after which the resistance is constant.

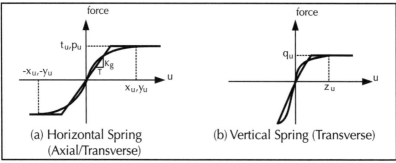

After ASCE, 1984

■ **Figure 5.2 Idealized Load-Deformation Relations at Pipe-soil Interface**

Relative movement parallel to the pipe axis results in longitudinal (horizontal axial) forces at the pipe-soil interface. For the elasto-plastic model, the ASCE guideline provides the following relations for clay and sand.

For sand,

$$t_u = \frac{\pi}{2} D \bar{\gamma} H (1 + k_o) tan\phi \qquad (5.1)$$

$$x_u = (0.1 \sim 0.2) \text{ in} = 2.54 \sim 5.08 \times 10^{-3} \text{ m} \qquad (5.2)$$

For clay,

$$t_u = \pi D \alpha S_u \qquad (5.3)$$

$$x_u = (0.2 \sim 0.4) \text{ in} = 5.08 \sim 10.16 \times 10^{-3} \text{ m} \qquad (5.4)$$

where D is the pipe diameter, S_u is the undrained shear strength of the surrounding soil, α is an empirical adhesion coefficient varying with S_u, $\bar{\gamma}$ is the effective unit weight of the soil, H is the depth to center-line of the pipeline, ϕ is the angle of shear resistance of the sand and k_o is the coefficient of lateral soil pressure at rest. The magnitude of k_o for normally consolidated cohesionless soil has been reported to range from 0.35 to 0.47. However, one expects k_o to be somewhat larger because of the backfilling and compaction of the soil around pipelines. T. O'Rourke et al. (1985) recommend that $k_o=1.0$, as a conservative estimate under most conditions of pipeline burial. Finally, k is the reduction factor depending on the outer-surface characteristics and hardness of the pipe. For a concrete pipe, steel or cast iron pipe with cement coating, $k=1.0$, for cast iron or rough steel, k ranges from 0.7 to 1.0, while for smooth steel or for a pipe with smooth, relatively hard coating, k ranges from 0.5 to 0.7.

The ASCE guideline recommends the relationship between α and S_u shown in Figure 5.3, in which the adhesion factor is a decreasing function of the undrained shear strength of the soil.

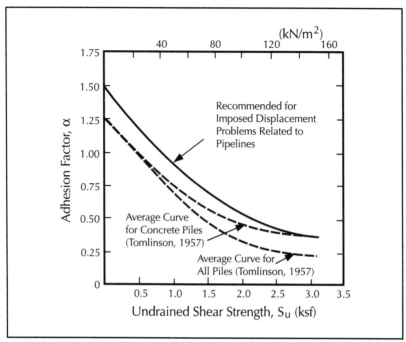

After ASCE, 1984

■ **Figure 5.3 Adhesion Factors vs. Undrained Shear Strength**

5.1.2 HORIZONTAL TRANSVERSE MOVEMENT

Relative movement perpendicular to the pipe axis in the horizontal plane results in horizontal transverse forces at the pipe-soil interface. For the elasto-plastic model, the ASCE guideline provides the following relations for sand and clay.

For sand,

$$p_u = \bar{\gamma} H N_{qh} D \qquad (5.5)$$

$$y_u = \begin{cases} (0.07 \sim 0.10)(H + D / 2) & \text{for loose sand} \\ (0.03 \sim 0.05)(H + D / 2) & \text{for medium sand} \\ (0.02 \sim 0.03)(H + D / 2) & \text{for dense sand} \end{cases} \qquad (5.6)$$

For clay,

$$p_u = S_u N_{ch} D \tag{5.7}$$

$$y_u = (0.03 \sim 0.05)(H + D/2) \tag{5.8}$$

where N_{qh}, N_{ch} are the horizontal bearing capacity factors for sand and clay, respectively. N_{qh} is shown in Figure 5.4 for sand. N_{ch} is equivalent to N_{qh} with $\phi=30°$.

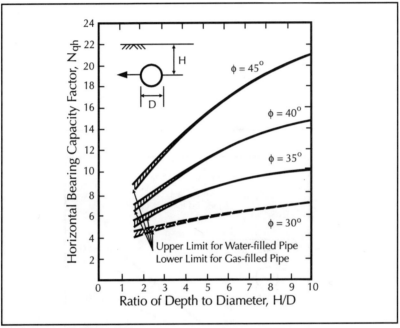

After Trautmann and T. O'Rourke, 1983

■ Figure 5.4 Horizontal Bearing Factor for Sand vs. Depth to Diameter Ratio

5.1.3 VERTICAL TRANSVERSE MOVEMENT, UPWARD DIRECTION

Relative upward movement perpendicular to the pipe axis results in lateral forces at the pipe-soil interface. For the elasto-plastic model, the ASCE guideline provides the following relations for clay and sand.

For sand,

$$q_u = \bar{\gamma} H N_{qv} D \qquad (5.9)$$

$$z_u = (0.01 \sim 0.015)H \qquad (5.10)$$

For clay,

$$q_u = S_u N_{cv} D \qquad (5.11)$$

$$z_u = (0.1 \sim 0.2)H \qquad (5.12)$$

where N_{qv} is the vertical uplift factor for sand and N_{cv} is the vertical uplift factor for clay.

N_{qv} and N_{cv} are presented as functions of the depth over diameter ratio in Figure 5.5 and Figure 5.6, respectively.

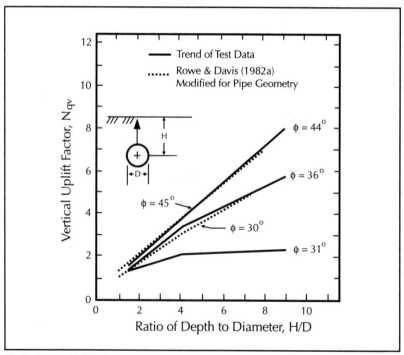

After Trautmann and T. O'Rourke, 1983

■ Figure 5.5 Vertical Uplift Factor for Sand vs. Depth to Diameter Ratio

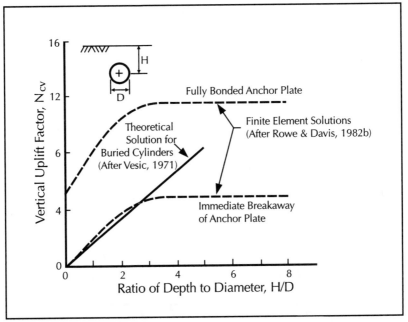

After ASCE, 1984

■ Figure 5.6 Vertical Uplift Factor for Clay vs. Depth to Diameter Ratio

5.1.4 VERTICAL TRANSVERSE MOVEMENT, DOWNWARD DIRECTION

Relative downward movement perpendicular to the pipe axis in the vertical plane results in lateral forces at the pipe-soil interface. For the elasto-plastic model, the ASCE guideline provides the following relations for sand and clay.

For sand,

$$q_u = \bar{\gamma} H N_q D + \frac{1}{2} \gamma D^2 N_y \qquad (5.13)$$

$$z_u = (0.10 \sim 0.15)D \qquad (5.14)$$

For clay,

$$q_u = S_u N_c D \qquad (5.15)$$

$$z_u = (0.10 \sim 0.15)D \qquad (5.16)$$

where γ is the total unit weight of sand, N_q and N_y are the bearing capacity factors for horizontal strip footings on sand loaded in the vertically downward direction, while N_c is the bearing capacity factor for horizontal strip footings on clay.

The ASCE guideline suggests that these three factors can be obtained from Figure 5.7.

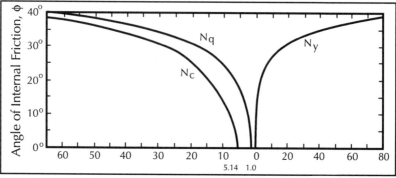

■ Figure 5.7 Vertical Bearing Capacity Factors vs. Soil Friction Angle

5.2

EQUIVALENT STIFFNESS OF SOIL SPRINGS

As noted previously, interaction at the soil-pipe interface can be modeled as an elastic spring as long as the relative displacement is less than the maximum elastic deformation x_u, etc. In such cases, pipeline response can be determined by "Beam on Elastic Foundation" types of analysis. However, these analyses apply only for small to moderate levels of ground deformation since the maximum elastic deformation (x_u, etc.) typically are small. For example, for a 0.3 m diameter pipe (12 in) in moderately dense sand ($\gamma=1.8\times10^{-3}$ kgf/cm^3, 112 pcf, $\phi=35°$) with $H=1.2$ m (3.9 ft), the maximum elastic deformation for longitudinal, transverse horizontal, transverse upwards and transverse downward are 3.8×10^{-3} m (0.15 in), 0.04 m (1.6 in), 0.018 m (0.71 in) and 0.038 m (1.5 in), respectively.

In the following subsection, alternate relations for these spring coefficients are compared with those from the ASCE guideline.

5.2.1 AXIAL MOVEMENT

For axial soil spring constant, the *Specifications for Seismic Design of High Pressure Gas Pipelines* (Japan, 1982) suggests that the soil spring constant is proportional to pipe diameter. M. O'Rourke and Wang (1978) suggest that the soil spring constant is twice the effective shear modulus of soil. Table 5.1 provides a comparison of these three approaches for a pipe with diameter of 0.15 (6 in), 0.30 (12 in) and 0.61 m (24 in) in moderate dense sand ($\phi=35°$, $\gamma=1.8\times10^{-3}$ kgf/cm^3, 112 pcf) and burial depth of 1.2 m (3.9 ft).

■ Table 5.1 Comparison of Axial Soil Spring Stiffness

Source	Formula	Stiffness (kgf/cm²)			Note
		D=6 in	D=12 in	D=24 in	
Japanese Gas Association	$\pi D K_o$	28.3	56.5	113	$K_o = 0.6$ kgf/cm^3
O'Rourke and Wang	$2G_s$	60.2	60.2	60.2	$G_s = 66.3\sqrt{\gamma H \dfrac{1+2k_o}{3}}$
ASCE Guideline (Equivalent)	$\dfrac{t_u}{x_u\ /\ 2}$	27.9	59.1	131	$x_u = 3.8\times10^{-3}$ m, $k_o=0.6$, $k=0.9$, parameters for Eq 5.1

As shown in Table 5.1, all three approaches match reasonably well for a 12-inch diameter pipe. However, because the M. O'Rourke and Wang approach does not depend on pipe diameter, it apparently underestimates the axial stiffness for a large-diameter pipe and overestimates the axial stiffness for a small-diameter pipe. The equivalent axial soil spring stiffness from the ASCE guideline matches well with that by the Japanese Gas Association for all three pipe sizes.

5.2.2 LATERAL MOVEMENT IN THE HORIZONTAL PLANE

For lateral movement in the horizontal plane, Audibert and Nyman (1977) proposed three soil spring constants as shown in

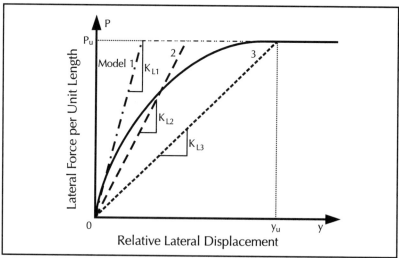

After Audibert and Nyman, 1977

■ Figure 5.8 Soil Spring Constants Corresponding to Different Relative Movement

Figure 5.8, for modeling the pipe-soil interaction as an elasto-plastic system. That is, K_{L1} and K_{L2} are for small and moderate relative displacements at the pipe-soil interface, respectively. K_{L3} is for relative displacements equal or large than y_u.

Similarly, Thomas (1978) suggests the following soil spring constant, K_L, for large lateral ground movement:

$$K_L = 2.7 \cdot \frac{p_u}{y_u} \qquad (5.17)$$

For small ground movement such as the movement induced by wave propagation, El Hmadi and M. O'Rourke (1989) suggest the following soil spring constant:

$$K_L = 6.67 \cdot \frac{p_u}{y_u} \qquad (5.18)$$

Note that these two constants are based on same interaction curve recommended by the ASCE guideline. Equation 5.18 corresponds to the initial slope of the interaction curve, K_{L1}, in Figure 5.8. Hence, this equation conservatively estimates the interaction force even for the wave propagation case.

5.2.3 VERTICAL MOVEMENT

Considering a infinite beam on an elastic subgrade, Vesic (1961) developed a soil spring constant using the Winkler hypothesis for downward pipe movement. The resulting vertical spring stiffness is:

$$K_v = 0.65 \left(\frac{E_s D^4}{E_p I_p} \right)^{\frac{1}{12}} \cdot \frac{E_s}{1 - \mu_s^2} \tag{5.19}$$

where E_s is the soil modulus, $E_s = 2(1+\mu_s)G_s$, μ_s is the Poisson ratio of soil, G_s is the shear modulus of soil, $E_p I_p$ is the flexural rigidity of the pipe.

For a steel pipe with a diameter of 30 cm (12 in) and a wall thickness of 0.76 cm (0.3 in), the vertical stiffness from Equation 5.19 ranges from 40 kgf/cm² to 260 kgf/cm² for G_s varying from 32 kgf/cm² to 600 kgf/cm². Using Equations 5.13 and 5.14 and assuming a moderately dense sand ($\phi=35°$, $\gamma=1.8\times10^{-3}$ kgf/cm³, 112 pcf) and burial depth of 1.2 m (3.9 ft), the ASCE guideline evaluates a soil spring constant of 140 kgf/cm² (2.0 kips/in²). Note that this value is about an average of the value from Equation 5.19.

5.3 LIQUEFIED SOIL

For a pipeline located in a liquefied layer as opposed to a competent layer, Suzuki et al. (1988) and Miyajima and Kitaura (1989) have shown that the pipe response is very sensitive to the stiffness of the equivalent soil springs. This subsection will discuss the equivalent stiffness of soil springs for a pipe in liquefied soil.

Combining experimental data with analytical solutions based on a beam on an elastic foundation approach, Takada et al. (1987) developed an equivalent soil spring for a pipe in a liquefied soil. They indicate that the equivalent stiffness ranges from 1/1000 to 1/3000 of that for non-liquefied soil. On the other hand, Yoshida and Uematsu (1978), Matsumoto et al. (1987), Yasuda et al. (1987), and Tanabe (1988) suggest that the stiffness ranges from 1/100 to 3/100 of that for non-liquefied soil based on their model experiments.

Miyajima and Kitaura (1991) also conducted model tests which indicated that the stiffness is related to the effective stress in the liquefied soil. That is, the soil spring constant is an increasing function of effective stress and hence, a decreasing function of excess pore water pressure ratio.

For saturated sandy soil, T. O'Rourke et al. (1994) proposed a reduction factor, R_f, for a pipe or pile subject to transverse ground displacement, as:

$$R_f = \frac{N_{qh}}{K_c} \cdot \frac{1}{0.0055(N_l)_{60}} \tag{5.20}$$

where K_c is the bearing capacity factor for undrained soil and $(N_l)_{60}$ is the corrected SPT value.

The reduced stiffness at the pipe-soil interface is then given by the stiffness for non-liquefied soil divided by the reduction factor. Their results suggest that the equivalent stiffness ranges from 1/100 to 5/100 of that for non-liquefied soil. Hence, both the longitudinal and transverse stiffnesses for a pipe in a liquefied soil can be taken as about 3% of the stiffness for a pipe in competent soil.

For the approach described above, the liquefied soil is treated, more or less, as a very soft solid. A liquefied soil can also be viewed as a viscous fluid. For that model, the interaction force at the pipe-soil interface varies with the relative velocity between the pipe and surrounding soil. According to Sato et al. (1994), the transverse force imposing on the pipe per unit length is:

$$F = \frac{4\pi\eta V}{2.002 - \log R_e} \tag{5.21}$$

where η is the coefficient of viscosity for the liquefied soil, V is the velocity of the pipe with respect to the liquefied soil, $R_e = \rho VD/\eta$, is the Reynolds number and ρ is the density of liquefied soil.

Based on model tests, Sato et al. (1994) established a relation between the coefficient of viscosity and the liquefaction intensity factor, F_L. This relation is shown in Figure 5.9.

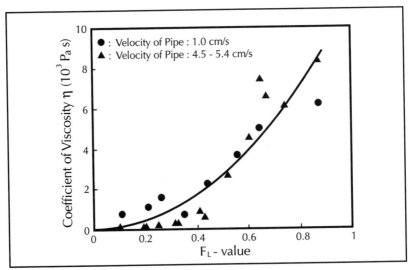

After Sato et al., 1994

■ Figure 5.9 Coefficient of Viscosity vs. F_L-Value

Japanese Road Association (1990) defined the factor of lique-faction intensity, F_L, as:

$$F_L = \frac{0.0042D_r}{(a_{max} / g) \cdot (\sigma_v / \sigma_v')}$$ (5.22)

where D_r is the relative density of the soil, a_{max} is the maximum acceleration of the ground, σ_v is the total overburden pressure and σ_v' is the effective overburden pressure.

Note that there are two problems with modeling liquified soil as a fluid. First of all, the velocity of the soil, which is an upper bound for the velocity of the soil with respect to the pipeline, typically is unknown. Secondly, when the liquefied soil stops flowing (i.e., V = 0), Equation 5.21 suggests that there would be no restoring spring force at the soil-pipeline interface which seems counterintuitive.

RESPONSE OF CONTINUOUS PIPELINES TO LONGITUDINAL PGD

As mentioned previously, PGD can be decomposed into longitudinal and transverse components. This chapter discusses the response of continuous pipeline subject to longitudinal PGD (soil movement parallel to the pipe axis). Subsequent chapters will cover the response of continuous pipeline to transverse PGD (soil movement perpendicular to the pipe axis) as well as the response of segmented pipe to PGD.

Under longitudinal PGD, a corrosion-free continuous pipeline may fail at welded joints, may buckle locally (wrinkle) in a compressive zone, and/or may rupture in a tensile zone. When the burial depth is very shallow, the pipeline may buckle like a beam in a ground compressive zone as discussed in Chapter 4.

Two separate models of buried pipe response to longitudinal PGD are presented herein. In the first model, the pipeline is assumed to be linear elastic. This model is often appropriate for buried pipe with slip joints since, as shown in Chapter 4, slip joints typically fail at load levels for which the rest of the pipe is linear elastic. In the second model, the pipeline is assumed to follow a Ramberg Osgood type stress-strain relation as given in Equation 4.1. This model is often appropriate for pipe with arc welded butt joints, since the local buckling or tensile rupture failure modes typically occur when the pipe is beyond the linear elastic range. Conditions leading to local buckling failure are presented, as well as those for tensile rupture. Finally the effect of flexible expansion joints are discussed. For each of these situations, case history comparisons are presented when available.

ELASTIC PIPE MODEL

As noted previously, the pattern of ground deformation has an effect upon the response of continuous pipelines to longitudinal PGD. Examples of observed longitudinal patterns were presented in Figure 2.10. For the purpose of analysis, M. O'Rourke and Nordberg (1992) have idealized five patterns as shown in Figure 6.1. That is, the Block pattern in Figure 6.1(a) is an idealization of the relatively uniform longitudinal pattern in Figure 2.10(a) (Section Line N-2) while the Ramp, Ramp-Block, Symmetric Ridge and Asymmetric Ridge pattern are idealization of the observed patterns in Figure 2-10 (b), (c), (d) and (e) respectively.

Assuming elastic pipe material and using either elasto-plastic or rigid-plastic force-deformation relations at the soil-pipe interface, M. O'Rourke and Nordberg (1992) analyzed the response of buried steel pipeline to three idealized patterns of longitudinal PGD (i.e., Ramp, Block and Symmetric Ridge).

Due to the small values for x_u, the maximum elastic deformation for the longitudinal "soil springs" discussed in Chapter 5, they found that the response for a simplified rigid-plastic model of the soil-pipe interaction gives essentially the same results as a more complex elasto-plastic model for the soil-pipe interface. The maximum pipe strain, ε, for all three patterns, normalized by the equivalent ground strain , is plotted as a function of the normalized length of the PGD zone in Figure 6.2. The length of the PGD zone is normalized by the embedment length, L_{em}, which is defined as the length over which the constant slippage force t_u must act to induce a pipe strain equal to the equivalent ground strain. Note that the Block pattern results in the largest strain in an elastic pipe.

For the Block pattern of PGD, the strain in an elastic pipe is then given by:

$$\varepsilon = \begin{cases} \dfrac{\alpha L}{2 L_{em}} & L < 4 L_{em} \\[2ex] \dfrac{\alpha L}{\sqrt{L L_{em}}} & L > 4 L_{em} \end{cases}$$

$$(6.1)$$

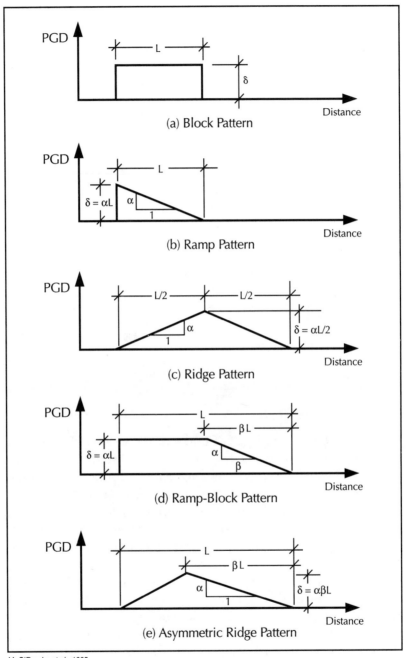

(a) Block Pattern

(b) Ramp Pattern

(c) Ridge Pattern

(d) Ramp-Block Pattern

(e) Asymmetric Ridge Pattern

M. O'Rourke et al., 1995

■ Figure 6.1 Five Idealization Patterns

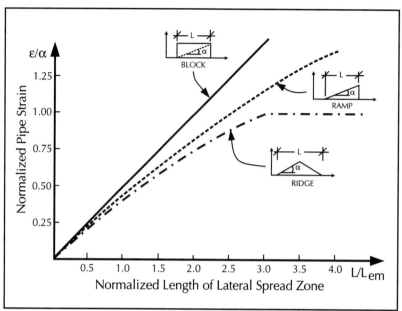

After M. O'Rourke and Nordberg, 1992

■ **Figure 6.2 Normalized Pipe Strain as Function of Normalized Length of the Lateral Spread Zone for Three Idealized Patterns of PGD**

$$L_{em} = \frac{\alpha EA}{t_u} \qquad (6.2)$$

Flores-Berrones and M. O'Rourke (1992) extended the model for a linear elastic pipe with a rigid-plastic "soil spring" (i.e., maximum resistance t_u for any non-zero relative displacement at the soil pipe interface) to the Ramp Block and Asymmetric Ridge patterns. They assigned the most appropriate of the five idealized patterns in Figure 6.1 to each of the 27 observed patterns presented by Hamada et al. (1986) and determined the peak pipe strain. They found that the idealized block pattern (that is Equation 6.1) gave a reasonable estimate of pipe response for all 27 of the observed patterns. This is shown in Figure 6.3 wherein the calculated maximum strain in two elastic pipes $\phi=27°$, $H=0.9$ m (3 ft), $t=1.9$ cm (3/4 in) for Pipe 1 and $\phi=35°$, $H=1.8$ m (6 ft), $t=1.27$ cm (1/2 in) for Pipe 2) with the appropriate idealized PGD pattern are plotted against the value from Equations 6.1 (i.e., an assumed Block pattern).

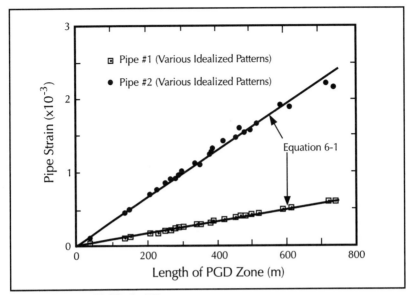

After Flores-Berrones and M. O'Rourke, 1992

■ **Figure 6.3 Maximum Strain in Two Elastic Pipes**

Some of the pipeline damage in the 1994 Northridge earthquake provide case histories for comparison with the elastic pipe model. Three out of seven pipelines along Balboa Blvd. were damaged due to the longitudinal PGD. Figure 6.4 shows a map of the PGD zone and the locations of the pipe breaks on Balboa Blvd.,

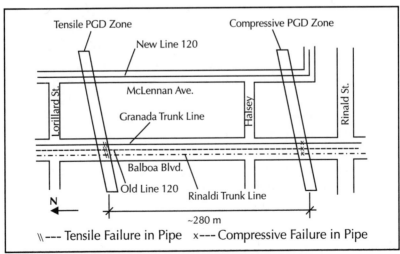

After M. O'Rourke and Liu, 1994

■ **Figure 6.4 Map of Ground Deformation Zones and Locations of Pipeline Damage on Balboa Blvd.**

in which the two parallelograms are the margins of the PGD zone. According to T. O'Rourke and M. O'Rourke (1995), the length or spatial extent of the PGD zone along Balboa Blvd. was 280 m (918 ft) and the amount of movement was about 0.50 m (20 in).

The properties of those three damaged pipelines are shown in Table 6.1. Note that all these pipelines are made of Grade-B steel.

■ **Table 6.1 Properties of Three Damaged Pipelines**

Pipeline	Diameter	Thickness	σ_y	Burial	Coating	Joints
	(mm)	(mm)	(MPa)	Depth (m)		
Granada Trunk Line	1260	6.4	249	1.8	Cement Mortar	Slip joints
Rinaldi Trunk Line	1730	9.5	249	2.7	Cement Mortar	Slip joints
Old Line 120	560	7.1	242	1.5	Coal Tar Epoxy	Unshielded Arc Welded

Due to the relatively low strength of slip joints and unshielded arc welded joints, the linear elastic pipeline model discussed above can be used to analyze the behavior of these pipelines.

The interaction force at the pipe-soil interface is evaluated using Equation 5.1, and assuming $\phi=37°$ for dense sand, $k=0.87$ for coal tar epoxy, $k=1.0$ for cement mortar coating and $\bar{\gamma} = 1.88 \times 10^4$ N / m^3 (115 pcf). For the two water trunk lines, the joint efficiency of 0.40 is assumed based upon Figure 4.9. For unshielded arc welded Line-120, the yield strain of the pipe steel is conservatively used as critical strain since the compressive and tensile strength of this type of joints are less than those determined from the yield strength of the pipe steel (T. O'Rourke and M. O'Rourke, 1995). Table 6.2 shows the critical strain for the pipes (i.e., failure condition based on joint efficiency, etc.) as well as the induced seismic strain calculated from Equation 6.1. Since the seismic strain is larger than the critical strain, failure is predicted for each of these pipes, which, as mentioned previously, was the observed behavior.

■ Table 6.2 Computation for Three Damaged Pipelines Along Balboa Blvd.

Pipeline	k	Frictional Coefficient	Joint Efficiency	Critical Strain	L_{em} (m)	Seismic Strain	Predicted Behavior
Granada Trunk Line	1.0	0.75	0.40	0.48×10^{-3}	50.0	2.1×10^{-3}	Failure
Rinaldi Trunk Line	1.0	0.75	0.40	0.48×10^{-3}	51.0	2.1×10^{-3}	Failure
Old Line 120	0.87	0.63	1.0	1.2×10^{-3}	78.7	1.59×10^{-3}	Failure

6.2

INELASTIC PIPE MODEL

As mentioned previously, local buckling of a pipeline with typical burial depths and arc-welded joints requires a model in which the pipe material is inelastic. Since the Block pattern appears to be the most appropriate model for elastic pipes, M. O'Rourke et al. (1995) assumed a Block pattern for determination of the circumstances leading to local buckling failure due to longitudinal PGD in a pipe with a more realistic Ramberg Osgood material model. The idealized Block pattern, shown in Figure 6.1(a), corresponds to a mass of soil having length L, moving down a slight incline. The soil displacement on either side of the PGD zone is zero, while the soil displacement within the zone is a constant value δ.

Two cases for a buried pipeline subject to a Block pattern of longitudinal PGD were considered. In Case I, the amount of ground movement, δ, is large and the pipe strain is controlled by the length, L, of the PGD zone. In Case II, L is large and the pipe strain is controlled by δ.

The distribution of pipe axial displacement, force and strain are shown in Figure 6.5 for Case I and in Figure 6.6 for Case II. Note that t_u is the friction force per unit length at the pipe-soil interface and L_e is the effective length over which t_u acts.

As shown in Figures 6.5 and 6.6, the force in the pipe over the segment AB is linearly proportional to the distance from Point A.

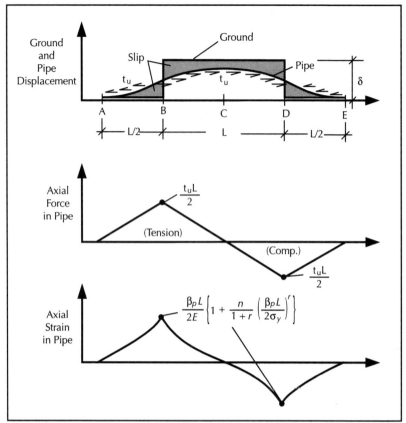

■ **Figure 6.5 Distribution of Pipe Axial Displacement, Force and Strain for Case I**

Using a Ramberg Osgood model, the pipe strain and displacement can be expressed as follows:

$$\varepsilon(x) = \frac{\beta_p x}{E}\left\{1 + \frac{n}{1+r}\left(\frac{\beta_p x}{\sigma_y}\right)^r\right\} \tag{6.3}$$

$$\delta(x) = \frac{\beta_p x^2}{E}\left\{1 + \frac{2}{2+r}\cdot\frac{n}{1+r}\cdot\left(\frac{\beta_p x}{\sigma_y}\right)^r\right\} \tag{6.4}$$

where n and r are Ramberg Osgood parameters discussed in Chapter 4, E is the modulus of elasticity of steel, σ_y is the effective yield stress and β_p is the pipe burial parameter, having units of pounds per cubic inch, defined below.

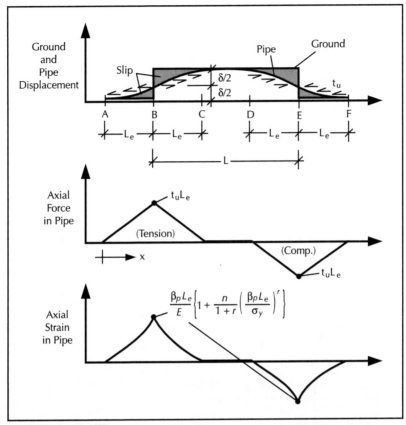

After M. O'Rourke et al., 1995

■ **Figure 6.6 Distribution of Pipe Axial Displacement, Force and Strain for Case II**

For sandy soil (c=0), the pipe burial parameter β_p is defined as:

$$\beta_p = \frac{\mu\gamma H}{t} \qquad (6.5)$$

where the frictional coefficient μ can be computed by:

$$\mu = \tan k\phi \qquad (6.6)$$

For clay, the pipe burial parameter β_p can be expressed as:

$$\beta_p = \frac{\alpha S_u}{t} \qquad (6.7)$$

6.2.1 WRINKLING

Critical displacement and spatial extent of the PGD zone can be determined by the equations discussed above. Substituting the critical local buckling strain into Equation 6.3, one can obtain the critical length of PGD zone L_{cr}. This can then be used to calculate the critical ground movement δ_{cr} from Equation 6.4. Using the Ramberg Osgood pipe material model, M. O'Rourke et al. (1995) develop critical values for δ and L which result in wrinkling of the pipe wall in compression (critical strain in compression taken as midpoint of range given in Equation 4.3). Table 6.3 shows these critical values for Grade-B ($n=10$, $r=100$) and X-70 ($n=5.5$, $r=16.6$) steel and a variety of burial parameters and R/t ratios (radius of pipe/thickness).

■ Table 6.3 Critical Length and Displacement for Compressive Failure of Grade-B and X-70 Steel and Various Burial Parameters and R/t Ratios

	R/t	$\beta_p = 1.0$ pci		$\beta_p = 2.5$ pci		$\beta_p = 5$ pci		$\beta_p = 15$ pci		$\beta_p = 25$ pci	
		L(m)	δ(m)	L(m)	δ(m)	L(m)	δ(m)	L(m)	δ(m)	L(m)	δ(m)
GR -B	10	1762	1.32	704	0.53	352	0.26	117	0.09	70	0.05
	25	1744	1.12	698	0.45	349	0.23	116	0.08	70	0.045
	50	1728	1.05	691	0.42	346	0.21	115	0.07	69	0.042
	100	1704	1.00	682	0.4	341	0.20	114	0.066	68	0.04
	150	1660	0.94	664	0.38	332	0.19	111	0.063	66	0.037
X- 70	10	4488	10.3	1795	4.1	898	2.10	299	0.69	180	0.41
	25	4182	6.87	1673	2.75	836	1.37	279	0.46	167	0.28
	50	3833	5.18	1533	2.1	768	1.04	256	0.35	153	0.21
	100	2577	2.25	1031	0.90	515	0.45	172	0.15	103	0.09
	150	1718	1.0	687	0.4	344	0.2	115	0.067	69	0.04

A pipe fails in local buckling when both the length and displacement of the PGD zone are larger than the critical values given, for example, in Table 6.3.

As a case history, two X-52 grade steel pipelines (Line 3000 and Mobil Oil) with arc welded joints subject to the longitudinal

PGD at Balboa Blvd. during the 1994 Northridge earthquake are studied here. These two pipelines are a 0.76 m diameter (30 in) gas pipeline and a 0.41 m diameter (16 in) oil pipeline, as listed in Table 6.4.

■ Table 6.4 Computation for Three Undamaged Pipelines Along Balboa Blvd.

Pipeline	D (m)	t (mm)	β_p (pci)	Compression		Tension	
				L_{cr} (m)	δ_{cr} (m)	L_{cr} (m)	δ_{cr} (m)
Line 3000	0.76	9.5	10	281	0.31	362	1.57
Mobil Oil	0.41	9.5	4	815	1.25	960	4.2

Based on the ASCE Guidelines (1984), the friction reduction factor is taken as 0.6 for a pipe with epoxy or polyethylene coatings. The corresponding burial parameter is 10 pci (0.28 kgf/cm^3) for the gas line and 4 pci (0.11 kgf/cm^3) for the Mobil line. The critical displacement δ_{cr} and the critical length L_{cr} for both wrinkling and tensile rupture were determined and listed in Table 6.4. For tensile rupture, the critical strain was taken as 4%. Since the calculated critical length is larger than the observed length of the PGD zone, the M. O'Rourke et al. (1995) model predicted successful the behavior of those two X-52 grade steel pipelines along Balboa Blvd. Note, however, that the procedure suggests that one of the lines (3000) is close to incipient wrinkling.

6.2.2 TENSILE FAILURE

As indicated previously, an initial compressive failure in steel pipes subject to longitudinal PGD is more likely than an initial tensile rupture failure. That is, since the peak pipe force and strain in tension and compression are equal as shown in Figure 6.5 (Case I) and 6.6 (Case II), and the critical failure strain in compression is less, one first expects a compressive failure. One can determine the conditions for an initial tensile rupture (for example in a pipe locally reinforced in the compression region near the toe of an expected lateral spread) by substituting a tensile rupture strain into

Equation 6.3. This gives the critical length of the longitudinal PGD zone which in combination with Equation 6.4 yields the critical ground displacement. Note that the critical parameters, L_{cr} and δ_{cr}, for tensile failure are larger than that for compression failure.

These values can be used to evaluate the likelihood of a subsequent tensile failure. By subsequent tensile failure, we mean a tensile rupture failure in the pipe near the head of the PGD zone, after a local buckling failure in the compression region near the toe. For Case II shown in Figure 6.6, the critical parameters can be used directly to determine the potential for a subsequent tensile rupture. In that case, L is large enough and δ is small enough so that a compression failure at Point E does not affect the state of stress in the tensile region around Point B, For Case I shown in Figure 6.5, the critical parameters for tensile failure are upper bounds for subsequent tensile failure since a compressive failure which limits the pipe force at Point D, increases the tensile force at Point B.

6.3

INFLUENCE OF EXPANSION JOINTS

M. O'Rourke and Liu (1994) studied the influence of flexible expansion joints in a continuous pipeline subject to longitudinal PGD. Depending upon the location of the expansion joints, they may have no effect, have a beneficial effect or have a detrimental effect. For example, referring to Figure 6.5 (Case I) if the expansion joint is located at a distance larger than L away from the center of the PGD zone (i.e., to the left of Point A or to the right of Point E in Figure 6.5), an expansion joint would have no effect on the pipe stress and strain induced by the longitudinal PGD since the axial force in the pipe would be zero there even with no expansion joints. Similarly for Case II shown in Figure 6.6, an expansion joint to the left of Point A, to the right of Point F, or between Points C and D would have no effect.

Figures 6.7 through 6.9 illustrate the beneficial effects of two expansion joints close to the head and toe areas of a longitudinal PGD zone. In all three cases an expansion joint is located within a distance $L_1(L_1<L/2)$ of the head of the PGD zone (Point B) and another within $L_2(L_2<L/2)$ of the toe (Point E). This placement is beneficial since the peak tension and compression forces are limited to $t_u L_1$ and $t_u L_2$ respectively.

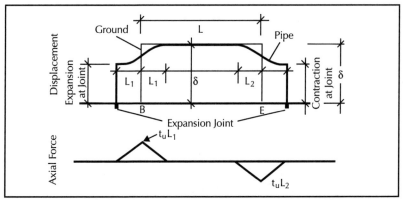

■ **Figure 6.7 Pipe and Soil Displacement with Two Expansion Joints Outside PGD Zone**

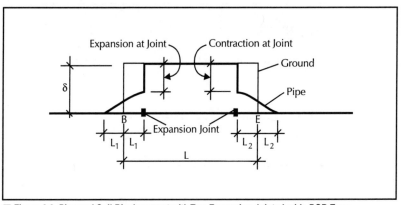

■ **Figure 6.8 Pipe and Soil Displacement with Two Expansion Joints Inside PGD Zone**

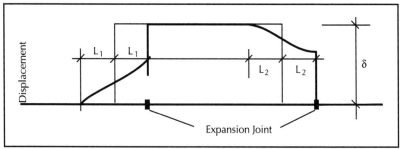

■ **Figure 6.9 Pipe and Soil Displacement with One Inside and One Outside PGD Zone**

Figure 6.10 illustrates the potential detrimental effects of a single expansion close to the head of a PGD zone.

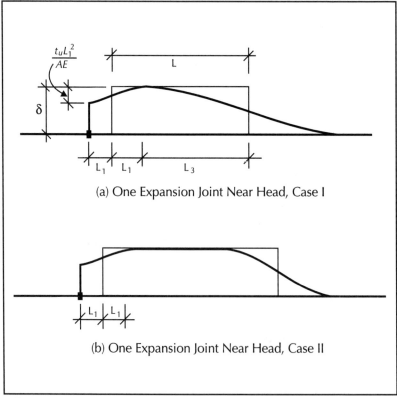

(a) One Expansion Joint Near Head, Case I

(b) One Expansion Joint Near Head, Case II

■ **Figure 6.10 Pipe and Soil Displacement with a Single Expansion Joint**

For Case I shown in Figure 6.10(a), the tensile stress is reduced to $t_u L_1$ but the compression stress is increased to $t_u L_3$. That is, a single expansion joint made the situation worse since the total load $t_u L$ is no longer shared equally at both the compression and tension zones. The reverse occurs for a single expansion joint near the toe region. That is for Case I the compression stress is reduced but the tensile stress increases. For Case II shown in Figure 6.10(b), the tensile stress is still reduced to $t_u L_1$ but the compression stress is unchanged.

The use of expansion joints presupposes that they are able to accommodate the imposed relative expansion and contraction. For example, if the distance L_1 and L_2 in Figure 6.9 are small (expansion joints very close to the head and toe of the PGD zone), the required expansion and contraction capability would essentially the same as the ground displacement δ. For an expansion joint at distance $L_1 (L_1 < L/2)$ away from the head or tow, the required expansion or contraction capacity is $\delta - t_u L_1^2/(AE)$ as shown in Figure 6.10.

If the amount of ground movement and required expansion/contraction capability are larger than the allowable movement of the expansion joint, the pipeline will likely be damaged. This type of damage has been observed in past events. For example, T. O'Rourke and Tawfik (1983) note that during the 1971 San Fernando earthquake, two water mains containing flexible joints were damaged at mechanical joints. However in that particular case, the pipe was subject to transverse PGD in combination of with a small amount of longitudinal PGD.

In summary, the use of expansion joints to mitigate against the effects of longitudinal PGD on continuous pipelines must be done with care. In general, to be effective, at least two expansion joints are needed, one close to the head of the PGD zone and the other close to the toe. In addition, the expansion and contraction capability of the joints themselves needs to be comparable to the amount of ground deformation δ. Finally, one needs a reasonably accurate estimate of both the location and extent of the PGD zone.

INFLUENCE OF AN ELBOW OR BEND

The response of a straight pipeline subject to longitudinal PGD is discussed in Sections 6.1 and 6.2. If an elbow or bend is located close to but beyond the margins of a longitudinal PGD zone, large pipe stresses may occur due to the induced bending moments. The case of a 90° bend in the horizontal plane is considered here. The longitudinal PGD causes the elbow to move in the direction of ground movement. This elbow movement is resisted by transverse soil springs along the transverse leg (i.e., the leg perpendicular to the direction of ground motion). The soil loading on the transverse leg results in bending moments, M, at the elbow as well as a concentrated force F (an axial force in the longitudinal leg and a corresponding shear force in the transverse leg). Similar to the models in Figures 6.5 and 6.6, two cases are considered herein as shown in Figure 6.11.

In both cases, the elbow is located at a distance L_o from the compression area (i.e., the ground movement is in a direction towards the elbow). In Case I, the length of the PGD zone, L, is small and the pipe response is controlled by the length of the PGD zone, as shown in Figure 6.11(a). In Case II, the length of the PGD zone is large and pipe response is controlled by the displacement of the PGD zone, δ, as shown in Figure 6.11(b).

In the analytical development which follows, the pipe is assumed to be elastic and the axial force per unit length along the longitudinal leg is taken as t_u. Assuming that the lateral soil spring along both the longitudinal and transverse legs are elastic, a beam on elastic foundation model of the elbow shown in Figure 6.12 results in the following relation between the imposed displacement δ' and the resulting force F:

$$\delta' = \frac{1}{2\lambda^3 EI}(F - \lambda M) \tag{6.8}$$

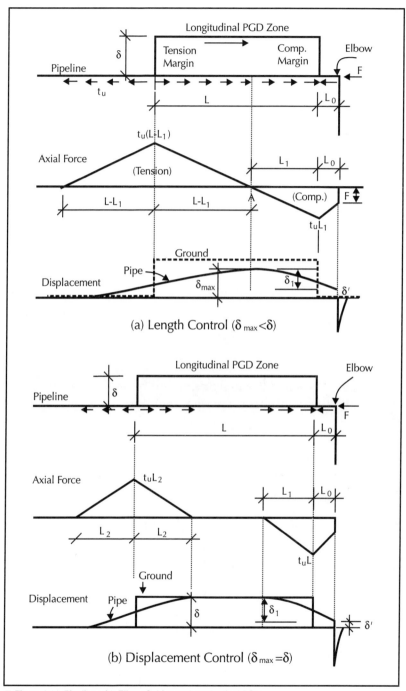

(a) Length Control ($\delta_{max} < \delta$)

(b) Displacement Control ($\delta_{max} = \delta$)

■ Figure 6.11 Pipeline with Elbow Subject to Longitudinal PGD

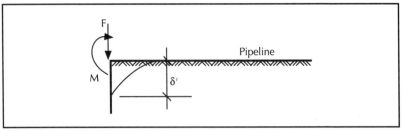

■ Figure 6.12 Model for Beam on Elastic Foundation

where,

$$\lambda = \sqrt[4]{\frac{k}{4EI}} \qquad (6.9)$$

For an imposed displacement δ' at the elbow, the relation between the resulting moment and force F is given by:

$$M = \frac{F}{3\lambda} \qquad (6.10)$$

For a given force F at the elbow, equilibrium of the longitudinal leg near the compression margin requires:

$$F = (L_1 - L_0)t_u \qquad (6.11)$$

where L_1 is the distance from the compression margin of the PGD zone to the point of zero axial pipe stress within the PGD zone (Point A in Figure 6.11(a)).

For a small L case (Case I in Figure 6.11(a)), the pipe response is controlled by the length of PGD zone. The maximum pipe displacement δ_{max} is less than the ground displacement δ. Considering pipe deformation near the compression margin,

$$\delta_{max} = \delta' + \delta_1 \qquad (6.12)$$

where δ' is given by Equations 6.8 and 6.10, and

$$\delta_1 = \frac{FL_0}{AE} + \frac{t_uL_0^2}{2AE} + \frac{t_uL_1^2}{2AE} \qquad (6.13)$$

is the displacement due to pipe strain between Point A and the elbow.

Considering pipe deformation near the tension margin, that is integrating pipe strain to the left of Point A gives,

$$\delta_{max} = \frac{t_u(L - L_1)^2}{AE} \tag{6.14}$$

hence

$$\frac{t_u(L - L_1)^2}{AE} = \frac{F}{3EI\lambda^3} + \frac{FL_0}{AE} + \frac{t_u L_0^2}{2AE} + \frac{t_u L_1^2}{2AE} \tag{6.15}$$

For a large length case (Case II in Figure 6.11(b)), the pipe response is controlled by the maximum ground displacement. That is,

$$\delta_{max} = \delta = \delta' + \delta_1 \tag{6.16}$$

or

$$\delta = \frac{F}{3EI\lambda^3} + \frac{FL_0}{AE} + \frac{t_u L_0^2}{2AE} + \frac{t_u L_1^2}{2AE} \tag{6.17}$$

The force F at the elbow and the effective length L_1 can be obtained by simultaneously solving Equations 6.11 and 6.15 (Case I) or Equations 6.11 and 6.17 (Case II). The moment can then be calculated by Equation 6.10. When the elbow is located near the compression margin, the maximum pipe stress at the elbow is then given by:

$$\sigma = \frac{F}{A} \pm \frac{MD}{2I} \tag{6.18}$$

When the elbow is located beyond the PGD zone but close to the tension margin, the same relation applies but the force at the elbow, F, would be tension.

As a case history, New Line 120 is considered herein. As shown in Figure 6.13, New Line 120 was subject to longitudinal PGD along McLennan Avenue during the 1994 Northridge earthquake. There is an elbow at about 40 m (131 ft) away from the southern (compression) margin of the PGD zone. The length of the PGD zone was about 280 m (918 ft) and the amount of ground displacement was reported to be about 0.50 m (20 in). This line is X-60 grade steel pipe with D=0.61 m (24 in), t=0.0064 m (1/4 in) and H=1.5 m (5.0 ft). For the pipe with fusion bounded epoxy coating (μ = 0.38), the axial friction force per unit length is estimated to be 1.8×10^4 N/m (103 lbs/in) from Equation 5.1 while the lateral (transverse) spring coefficient is estimated to be 1.67×10^6 N/m^2 (242 lbs/in^2) (1.0×10^5 N/m (571 lbs/in) from Equation 5.5, divided by 0.06 m (2.4 in) from Equation 5.6).

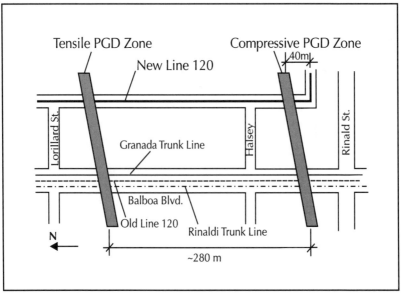

■ Figure 6.13 Observed PGD and Location of Pipeline

For Case I, Equations 6.11 and 6.15 suggest that L_i=90 m (295 ft) and Equation 6.13 gives δ_{max}=0.253 m (10 in) while for Case II Equations 6.11 and 6.17 suggest L_i=110 m (361 ft) and δ_{max}=0.50 m (20 in). Based on these parameters, the case history corresponds to Case I (i.e., length control).

For L_i=90 m (295 ft), the peak pipe stress at the compression margin $t_u L_i/A$=132 MPa (19.1 ksi), the axial stress at the elbow F/A=73 MPa (10.6 ksi) while the bending stress at the elbow M/S=1100 MPa (159 ksi). Hence, the total stress at the elbow, assuming the pipe remains elastic, is 1173 MPa (170 ksi). Note that these values, based on the analytical relation for an elastic pipe compare favorably with corresponding FE results for an elastic pipe.

Note that the results presented here are based on an elastic assumption for both the lateral soil spring coefficients and pipe material. For non-linear soil spring coefficients and elasto-plastic pipe material, the maximum stresses at the bend are less. For the New Line 120, a finite element analysis shows that the compressive stress at the compression margin is 517 MPa (75 ksi). The maximum tensile strain is 2.78×10^{-3} while the maximum compressive strain is 4.02×10^{-3} at the elbow. This compressive strain is larger than the critical strain for local buckling (1.80×10^{-3} from Equation 4.3). Hence, wrinkling is expected at the bend, but not at the compression margin.

The authors understand that the line was inspected after the Northridge event and is currently in service. No distress was noted at either the tension or compression margin. However, the elbow region was not inspected.

RESPONSE OF
CONTINUOUS PIPELINES
TO TRANSVERSE PGD

As mentioned previously, transverse PGD refers to permanent ground movement perpendicular to the pipe axis. When subject to transverse PGD, a continuous pipeline will stretch and bend as it attempts to accommodate the transverse ground movement. The failure mode for the pipe depends then upon the relative amount of axial tension (stretching due to arc length effects) and flexural (bending) strain. That is, if the axial tension strain is low, the pipe wall may buckle in compression due to excessive bending. On the other hand, if axial tension is not small, the pipe may rupture in tension due to the combined effects of axial tension and flexure. T. O'Rourke and Tawfik (1983) present a case history, from the 1971 San Fernando event, of continuous pipe failure due to PGD. The transverse component of PGD was approximately 1.7 m. Line 1001 (Pipeline 5 in Figure 2.11 (a)) was abandoned because of multiple breaks. Line 85 (Pipeline 4 in Figure 2.11(a)) was repaired at several locations within the PGD zone. The records indicate that three repairs near the eastern boundary of the soil movement were due to tensile failure and two other repairs near the western boundary were due to compressive failure. Note that besides the major lateral movement, there was small axial movement toward the west.

Similar to longitudinal PGD, pipeline response to transverse PGD is in general a function of the amount of PGD δ, the width of the PGD zone as well as the pattern of ground deformation. Figure 7.1 presents sketches of two types of transverse patterns considered herein.

Observed examples of **spatially distributed** transverse PGD (sketched in Figure 7.1(a)) have previously been presented in Figure 2.11(b) and (c) and in Figure 2.11(a) near Pipeline 2. In these cases, the pipe strain is a function of both the amount and width of the PGD zone. Observed examples of **abrupt** transverse PGD (sketched in Figure 7.1(b)) have previously been presented in Figure 2.11(a) near Pipelines 4 and 5. In these cases, the pipe strain is

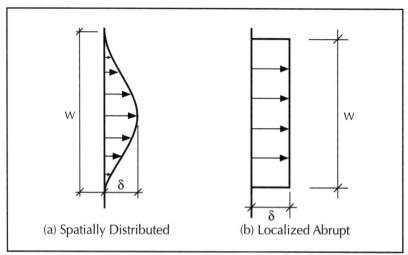

(a) Spatially Distributed   (b) Localized Abrupt

■ Figure 7.1 Patterns of Transverse PGD

a function of δ and in some cases, the width of the zone W. That is, if the zone is wide, the movement at each margin of the PGD zone corresponds more or less to a fault offset where the fault/pipeline intersection angle is 90°.

Another type of transverse PGD occurs when a pipe is buried directly in liquefied soil. In addition to the pipe deformation in the horizontal direction due to lateral spreading of liquefied soil, it may also uplift due to buoyancy (transverse deformation in the vertical direction). This mechanism has caused pipe damage in past events. For example, Suzuki (1988) and Takada (1991) mentioned that some pipes, with or without manholes, were uplifted out of the ground due to buoyancy effects during the 1964 Niigata earthquake.

In this chapter, we discuss continuous pipeline response to spatially distributed transverse PGD in detail. Various analytical idealizations of transverse PGD which have been used are reviewed. Analytical and numerical models of pipe response to spatially distributed transverse PGD are then discussed. The cases of a pipeline in a competent soil layer and in a liquefied layer are presented separately. Also, the effects of the buoyance force are discussed. Finally, the conditions whereby abrupt transverse PGD may be modeled as a special case of fault offset are presented. The topic of pipe response to fault offset is discussed in Chapter 8.

IDEALIZATION OF SPATIALLY DISTRIBUTED TRANSVERSE PGD

One of the first items needed to evaluate pipeline response to spatially distributed transverse PGD is the pattern of ground deformation, that is, the variation of ground displacement across the width of the PGD zone. Different researchers have used different patterns in their analyses.

T. O'Rourke (1988) approximates the soil deformation with the beta probability density function.

$$y(x) = \delta[s/s_m]^{r'-1}[(1 - s)/(1 - s_m)]^{\tau - r' - 1} \quad 0 < s < 1 \qquad (7.1)$$

where s is the distance between the two margins of the PGD zone normalized by the width W, s_m is the normalized distance from the margin of the PGD zone to the location of the peak transverse ground displacement, δ, while r' and τ are parameters of the distribution. In his analysis, the following values were used; $s_m = 0.5$, $r' = 2.5$ and $\tau = 5.0$. Figure 7.2 shows the resulting idealized soil deformation.

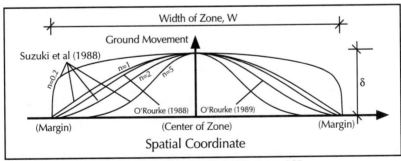

■ Figure 7.2 Assumed Patterns for Spatially Distributed Transverse PGD

Suzuki et al. (1988) and Kobayashi et al. (1989) approximate the transverse soil deformation by a cosine function raised to a power n.

$$y(x) = \delta \cdot \left(\cos \frac{\pi x}{W} \right)^n \tag{7.2}$$

where the non-normalized distance x is measured from the center of the PGD zone. Figure 7.2 also shows the Suzuki et al. and Kobayashi et al. model for $n = 0.2$, 1.0, 2.0 and 5.0.

M. O'Rourke (1989) assumes the following function for spatially distributed transverse PGD:

$$y(x) = \frac{\delta}{2} \left(1 - \cos \frac{2\pi x}{W} \right) \tag{7.3}$$

where x is the non-normalized distance from the margin of the PGD zone. This gives the same shape as both the Suzuki and Kobayashi et al.'s models with $n = 2.0$ (note, origin of x axis is shifted).

As shown in Figure 7.2, all the patterns are similar in that the maximum soil deformation occurs at the center of the PGD zone and the soil deformation at the margins is zero. The patterns differ in the variation of ground deformation between the center and the margins.

7.2 PIPELINE SURROUNDED BY NON-LIQUEFIED SOIL

Pipelines are typically buried about 1.0 m (3 ft) below the ground surface. Often the ground water level and the top surface of the liquefied soil layer are both below the bottom of the pipe. In these cases the force-deformation relations at the soil-pipeline interface correspond to a pipe in competent non-liquefied soil which overrides a liquefied soil layer.

In the following subsections, results from various analytical approaches and nonlinear finite element approaches will be presented and compared. Results for pipes in liquefied soil are presented in Section 7.3.

7.2.1   FINITE ELEMENT METHODS

The finite element method allows explicit consideration of the non-linear characteristics of pipe-soil interaction in both the transverse and longitudinal direction as well as non-linear stress-strain relations for pipe material. T. O'Rourke (1988), Suzuki et al. (1988) and Kobayashi et al. (1989) as well as Liu and M. O'Rourke (1997b) have used the finite element approach to evaluate buried pipe response to spatially distributed transverse PGD. Assumptions and numerical results from each group are presented here.

T. O'Rourke

T. O'Rourke (1988) simulated the soil deformation by the beta probability density function given in Equation 7.1. Figure 7.3 shows the deformation of both the soil and the pipe.

As shown in Figure 7.3, L_a is the distance from the margin of the PGD zone to an assumed anchored point in the undisturbed soil beyond the PGD zone. The anchored point in the T. O'Rourke (1988) model was located where the bending strain is less than 1×10^{-5}.

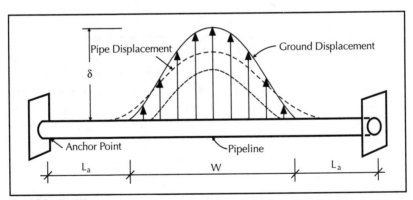

After T. O'Rourke, 1988

■ Figure 7.3 Parameters for T. O'Rourke's Model

Figure 7.4 presents the maximum tensile strain versus the maximum ground displacement for various widths of the PGD zone for an X-60 pipe with 0.61 m (24 in) diameter, 0.0095 m (3/8 in) wall thickness and burial depth $H = 1.5$ m (5 ft). For the three widths considered, as shown in Figure 7.4, the width of 10 m (33 ft) results in the largest tensile strain in the pipe for any given value of δ.

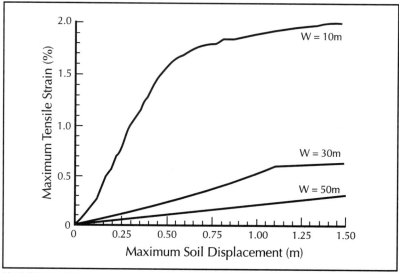

After T. O'Rourke, 1988

■ **Figure 7.4 Maximum Tensile Strain vs. Maximum Ground Displacement for Various Width of PGD Zone**

Figure 7.5 presents the maximum compressive strain as a function of δ for a width of 30 m. In this plot, the soil density ranged from 18.8 to 20.4 kN/m³ (115 to 122 pcf) and the soil friction angle ranged from 35° to 45°. Note there is no difference in pipe response for δ < 0.5 m (1.6 ft) and only a 30% difference for δ = 1.5 m (5 ft). Based on these observations, T. O'Rourke (1988) concluded that the width of the PGD zone has a greater influence on the magnitude of pipe strains than the soil properties.

From Figures 7.4 and 7.5, the peak tensile and compressive strains for a width of 30 m (98 ft) and δ = 1.5 m (5 ft) are about 0.61% and 0.32%, respectively. This indicates that the induced axial pipe strain at least in the T. O'Rourke (1988) model, is significant.

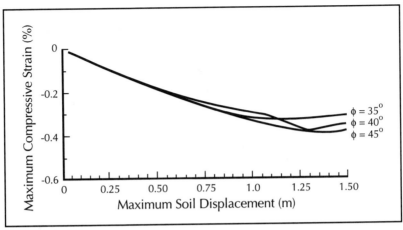

After T. O'Rourke, 1988

■ Figure 7.5 Maximum Compressive Strains vs. Maximum Ground Displacement for Different Soil Friction Angles

Suzuki et al.

Suzuki et al. (1988) expressed the pattern of transverse ground displacements by the cosine function raised to the n power as given in Equation 7.2. The normalized patterns for four values of n are shown in Figure 7.2. The patterns for n close to zero approximate abrupt transverse PGD while the patterns for $n \geq 1$ correspond to spatially distributed transverse PGD.

Suzuki et al.'s physical model is similar to T. O'Rourke's shown in Figure 7.3 except for the PGD pattern and the anchored length L_a. Suzuki et al. note that L_a needs be long enough such that the axial friction at the pipe-soil interface can fully accommodate the axial movement of the pipe due to the PGD. That is, there should be no flexural or axial strain in the pipe at the anchor points. It turns out that the anchored length in Suzuki et al.'s model is much larger than that in the T. O'Rourke (1988) model.

Figure 7.6 presents the influence of the width of the PGD zone on pipe strain for X-52 grade steel, 0.61 m (24 in) diameter, 0.0127 m (1/2 in) wall thickness and $H = 1.5$ m (5 ft). For given values of W and δ, the tensile and compressive strains are about equal. This suggests that the axial strain in the pipe is small. A certain width of the PGD zone somewhere around 30 m (98 ft) results in the largest pipe strain. Note that although the pipes are somewhat different, the tensile pipe strains in Figures 7.4 and 7.6 are similar for $W = 30$ and 50 m (164 ft).

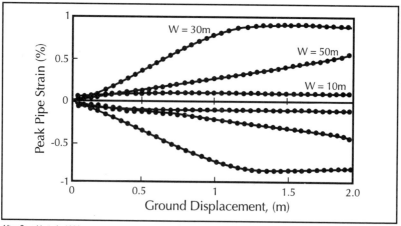

After Suzuki et al., 1988

■ **Figure 7.6 Maximum Strain vs. PGD for Different Width of the PGD Zone; X-52 Grade Steel**

Kobayashi et al.

Kobayashi et al. (1989) used the same shape function and followed the same procedure as Suzuki et al.'s to deal with the problem. They consider an X-42 grade steel pipe with 0.61 m (24 in) diameter and 0.0095 m (3/8 in) wall thickness. Kobayashi et al.'s results for the peak tensile strain are shown in Figure 7.7 for various of widths of the PGD zone. Note that the largest pipe strain occurs for a width of about 19 m (62 ft) in their model.

Liu and M. O'Rourke

Liu and M. O'Rourke (1997b) developed a finite element model, utilizing large deformation theory, non-linear pipe-soil interaction forces (soil springs) and Ramberg Osgood stress-strain relations for the pipe material. The pipe is modeled as a beam coupled by both axial and lateral soil springs. The anchor length of the pipe is long enough (up to 400 m (1312 ft)) such that both the flexural and axial pipe strain are essentially zero at the two anchor points. The pipe is assumed surrounded by loose to moderately dense sand (friction angle $\phi = 35°$ and soil density $\gamma = 1.87 \times 10^4 \, N/m^3$ (115 pcf)) with a burial depth $H_c = 1.2$ m (4 ft) from ground surface to the top of the pipe. The resulting elasto-plastic soil springs are based on the TCLEE Guideline (ASCE, 1984) and

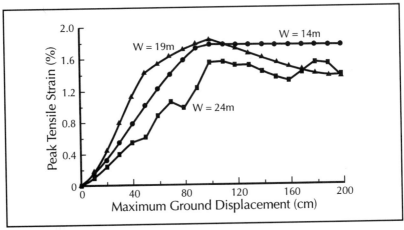

After Kobayashi et al., 1989

■ **Figure 7.7 Peak Tensile Strain vs. Maximum Ground Displacement; X-42 Grade Steel**

have peak transverse, p_u, and longitudinal, t_u, resistance of 1.0×10^5 and 2.4×10^4 N/m (571 and 137 lbs/in) respectively. The relative displacements between pipe and soil at which the peak transverse and longitudinal soil resistances are mobilizing are 0.06 and 3.8×10^{-3} (2.4 and 0.15 in), respectively.

Figure 7.8 shows the maximum tensile and compressive strains in the pipe versus the ground displacement for $W = 10$, 30 and 50 m, while Figure 7.9 shows the maximum pipe displacement versus the maximum ground displacement. Both these figures are for

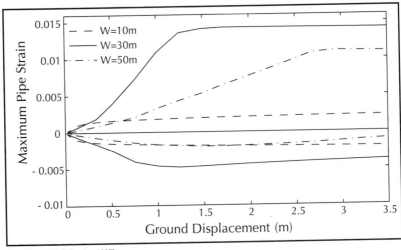

After Liu and M. O'Rourke, 1997b

■ **Figure 7.8 Maximum Pipe Strain vs. Ground Deformation; X-52 Grade Steel** an X-52 grade steel pipe with $D = 0.61$ m (24 in), $t = 0.0095$ m (3/8 in) and the ground deformation pattern given in Equation 7.3. Except for $W = 10$ m, Figure 7.8 indicates that the peak tensile strain is substantially larger than the peak compressive strain, particularly for larger values of δ. Also, for the three widths considered, the pipe strains are largest for $W = 30$ m. Although the pipes are somewhat different, the peak tensile strains shown in Figure 7.8 match reasonably well with Suzuki et al.'s shown in Figure 7.6 for all three widths. Also, both the peak tensile and compressive strains match reasonably well with the T. O'Rourke (1988) results for $W = 30$ and 50 m.

As shown in Figure 7.9, the maximum pipe displacement more or less matches the ground deformation up to a certain critical displacement δ_{cr}. Thereafter, the pipe strain remains relatively constant while the pipe displacement increases more slowly with ground deformation. For ground deformation greater than δ_{cr}, the pipe bending strain varies slightly (increasing for small widths and decreasing for large widths) and axial strain increases slowly, which results in the maximum tensile strain remaining more or less constant.

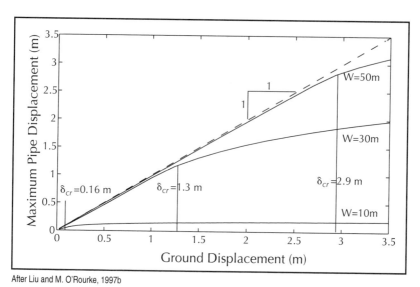

After Liu and M. O'Rourke, 1997b

■ Figure 7.9 Maximum Pipe Displacement vs. Ground Displacement; X-52 Grade Steel

For a fixed value of the width of the PGD zone ($W = 30$ m), Figure 7.10 shows the spatial distribution of pipe and soil displacement for $\delta = 0.5\delta_{cr}$, δ_{cr} and $2\delta_{cr}$.

Note that the pipe deformation matches fairly well with the ground deformation over the whole width of the PGD zone for $\delta \leq \delta_{cr}$. However, for $\delta > \delta_{cr}$, the maximum pipe displacement is less than the maximum ground displacement (from Figure 7.10, 40% less for $\delta = 2\delta_{cr}$), and "width" of the deformed pipe (i.e., length over which the pipe has noticeable transverse displacement) is larger than the width of the PGD zone. As a result, the curvature of the pipe is substantially less than the curvature of the ground for $\delta > \delta_{cr}$. As shown in Figure 7.10 for $W = 30$ m, the pipe curvature at $\delta = 2\delta_{cr}$ is comparable to the pipe curvature at $\delta = \delta_{cr}$.

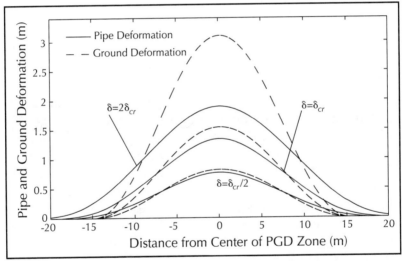

After Liu and M. O'Rourke, 1997b

■ **Figure 7.10 Pipe and Ground Deformation for W = 30 m; X-52 Grade Steel**

Figures 7.11 and 7.12 show the distribution of bending moments and axial forces in the pipe at $\delta = \delta_{cr}$ for $W = 10$, 30 and 50 m. As one might expect, the bending moments in Figure 7.11 are symmetric with respect to the center of the PGD zone and similar to those for a laterally loaded beam with built-in (i.e., fixed) supports near the margins of the PGD zone. That is, there are positive moments near the center of the PGD zone and negative moments

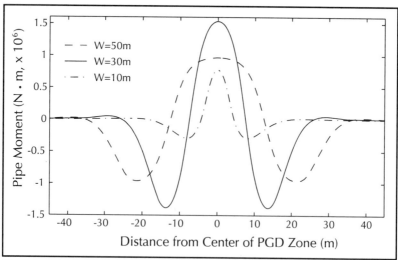

After Liu and M. O'Rourke, 1997b

■ **Figure 7.11 Distribution of Bending Moment for Three Widths ($\delta = \delta_{cr}$)**

near the margins. The moments vanish roughly 10 m beyond the margins. Note that the bending moments for $W = 30$ m are larger than those for $W = 10$ m or 50 m.

The axial forces in the pipe shown in Figure 7.12 are, as expected, also symmetric about the center of the PGD zone. The axial forces are maximum near the center of the zone and decrease in a fairly linear fashion with increasing distance from the center of the zone. Unlike the moments, the axial forces become small only at substantial distances beyond the margins of the zone (note the different distance scales in Figures 7.11 and 7.12). Also, for the three widths considered, the axial force was an increasing function of the width of the PGD zone (i.e., largest for $W = 50$ m and smallest for $W = 10$ m).

The transverse loading on the pipe also results in axial movement of the pipe, that is, inward movement towards the center of the PGD zone. This inward movement is an increasing function of the ground movement δ as shown in Figure 7.13. For $\delta = 4$ m (13 ft), this inward movement at the margins of the PGD zone for the pipe under consideration was 0.002, 0.07 and 0.15 m (0.08, 2.8 and 5.9 in) respectively for $W = 10$, 30 and 50 m.

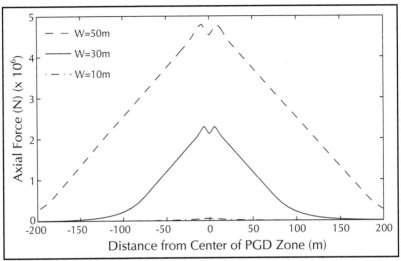

After Liu and M. O'Rourke, 1997b

■ **Figure 7.12** **Distribution of Axial Force for Three Widths ($\delta = \delta_{cr}$)**

The influence of other parameters upon the pipe behavior was also determined and is shown in Figures 7.14 through 7.20. Unless otherwise indicated, these results are for W = 30 m, X-52 grade steel, D = 0.61 m (24 in), t = 0.0095 m (3/8 in), p_u = 1.0×10^5

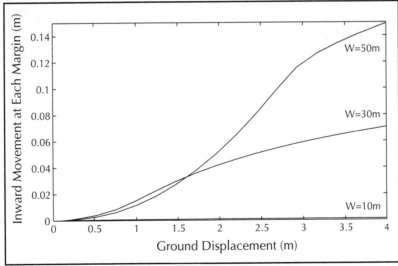

■ **Figure 7.13** **Pipe Inward Movement at Each Margin of PGD Zone**

N/m, (571 lbs/in), $t_u = 2.4 \times 10^4$ N/m (137 lbs/in) and the M. O'Rourke (1989) pattern of ground deformation. Figure 7.14 shows, for example, the influence of diameter on peak tensile and compressive strains. Note that both the peak tensile and compressive strains are increasing functions of diameter.

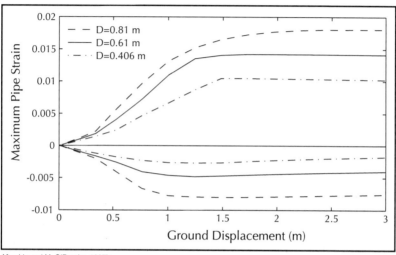

After Liu and M. O'Rourke, 1997b

■ Figure 7.14 Influence of Pipe Diameter, D

For the pipe model considered the peak tensile strain is, to a greater or lesser extent, a function of all the parameters shown in Figures 7.14 through 7.20. However, the peak compressive strain is essentially independent of the wall thickness, as shown in Figure 7.15, and the steel grade, as shown in Figure 7.18.

The peak tensile strain is an increasing function of the pipe diameter and the transverse (lateral) soil spring resistance. It is a decreasing function of the pipe wall thickness, the steel grade and to a lesser extent the longitudinal (axial) soil spring resistance.

In terms of anchor length L_a, a zero anchor length resulted in substantially larger pipe strain than $L_a = 15$ m (49 ft) or 400 m (1312 ft) as shown in Figure 7.19.

With reference to the PGD pattern, the M. O'Rourke (1989) pattern (same as the Suzuki et al. pattern with $n = 2$) resulted in the largest pipe strain for $W \geq 30$ m as shown in Figure 7.20. However, the Suzuki et al. (1988) pattern with $n = 1$ resulted in the largest strain for $W \leq 10$ m.

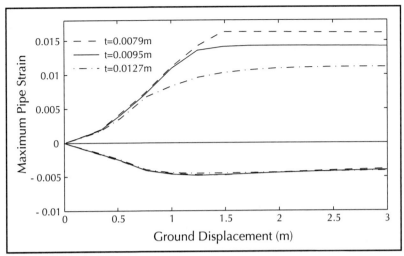

After Liu and M. O'Rourke, 1997b

■ Figure 7.15 Influence of Wall Thickness, t

The parameter which most strongly influences the tensile strain is the width of the PGD zone, followed by the transverse soil spring resistance, pipe diameter, steel grade, wall thickness, PGD pattern, anchor length of the pipe and longitudinal soil spring

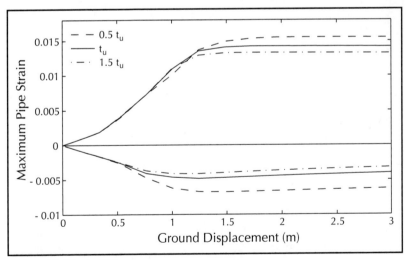

After Liu and M. O'Rourke, 1997b

■ Figure 7.16 Influence of Peak Longitudinal Soil Resistance, t_u

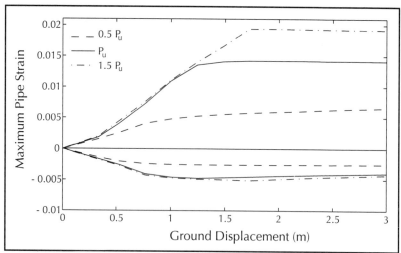

After Liu and M. O'Rourke, 1997b

■ **Figure 7.17 Influence of Peak Transverse Soil Resistance, p$_u$**

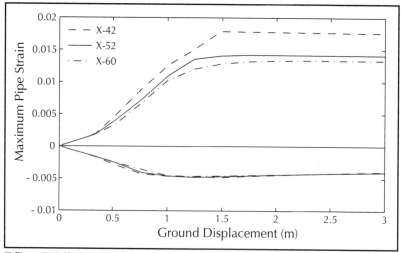

■ Figure 7.18 Maximum Pipe Strain for Three X-Grade Steel Materials

resistance. The critical ground displacement δ_{cr} was found to be an increasing function of width of the PGD zone and the lateral pipe-soil interaction force, but a decreasing function of steel grade, pipe diameter, axial pipe-soil interaction force and pipe wall thickness.

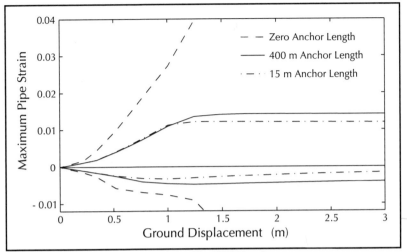

■ Figure 7.19 Effect of Anchored Length Outside of PGD Zone L$_a$

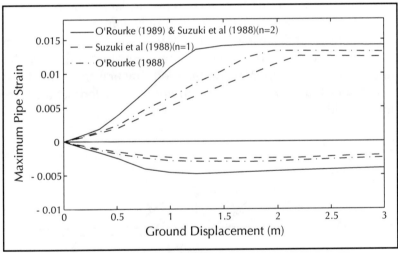

■ Figure 7.20 Effect of Spatially Distributed PGD Patterns

7.2.2 ANALYTICAL METHODS

Miyajima and Kitaura

Miyajima and Kitaura (1989) model a pipe subject to spatially distributed transverse PGD as a beam on an elastic foundation as shown in Figure 7.21. The equilibrium equations for the pipe are expressed as follows:

$$EI\frac{d^4y_1}{dx^4} + K_1 y_1 = K_1 \delta\left(1 - \sin\frac{\pi x}{W}\right) \quad \left(0 < x < \frac{W}{2}\right) \quad (7.4)$$

$$EI\frac{d^4y_2}{dx^4} + K_2 y_2 = 0 \quad \left(x \geq \frac{W}{2}\right) \quad (7.5)$$

where y_1 and y_2 are the transverse pipe displacement in and outside the PGD zone, K_1 and K_2 are the equivalent lateral soil spring coefficient in and outside the PGD zone, and EI is the flexural rigidity of the pipe cross-section. The equivalent soil springs K_1 and K_2 are based upon recommended practice in Japan (Japan Gas Association, 1982) in which non-linear characteristics are taken into consideration.

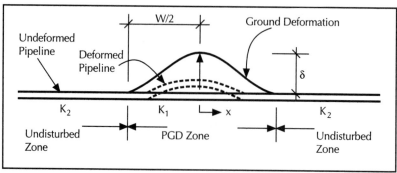

After Miyajima and Kitaura, 1989

■ Figure 7.21 Analytical Model for Pipeline Subject to Spatially Distributed Transverse PGD

Miyajima and Kitaura's equations provide a clear mechanical model and are solved by using a modified transfer matrix method. The maximum bending stress for a 16 inch (40 cm) diameter and

1/4 inch (0.6 cm) wall thickness steel pipe in a competent soil layer above the liquefied layer (i.e., $K_1 = K_2$) is shown in Figure 7.22 as a function of the width of the PGD zone W for three values of ground deformation δ.

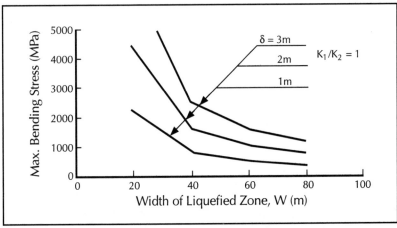

After Miyajima and Kitaura, 1989

■ Figure 7.22 Maximum Bending Stress vs. Width of PGD Zone for Three Values of Ground Deformation

As one might expect intuitively, the pipe stress is an increasing function of the ground deformation δ. For a given value of δ, the stress is a decreasing function of W for the range of widths considered by Miyajima and Kitaura. Note that they used small deformation flexural theory, which does not account for axial strain due to arc-length effects.

M. O'Rourke

M. O'Rourke (1989) developed a simple analytical model for pipeline response to spatially distributed transverse PGD. He considered two types of response as shown in Figure 7.23. For a wide width of the PGD zone, the pipeline is relatively flexible and its lateral displacement is assumed to closely match that of the soil. For this case, the pipe strain was assumed to be mainly due to the ground curvature (i.e., displacement controlled). For a narrow width, the pipeline is relatively stiff and the pipe lateral displace-

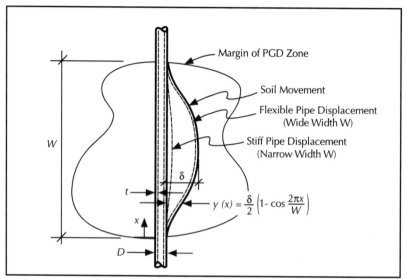

After M. O'Rourke, 1989

■ Figure 7.23 M. O'Rourke's Analytical Model for Pipeline Subject to Spatially Distributed Transverse PGD

ment is substantially less than that of the soil. In this case, the pipe strain was assumed to be due to loading at the soil-pipe interface (i.e., loading controlled).

For the wide PGD width/flexible pipe case, the pipe is assumed to match the soil deformation given by Equation 7.3. The maximum bending strain, ε_b, in the pipe, is given by:

$$\varepsilon_b = \pm \frac{\pi^2 \delta D}{W^2} \tag{7.6}$$

In this simple model, the axial tensile strain is based solely upon the arc-length of the pipe between the PGD zone margins. Assuming the pipe matches exactly the lateral soil displacement, the average axial tensile strain, ε_a, is approximated by:

$$\varepsilon_a = \left(\frac{\pi}{2}\right)^2 \left(\frac{\delta}{W}\right)^2 \tag{7.7}$$

For the narrow width/stiff pipe case, the pipe is modeled as a beam, built-in at each margin (i.e., fixed-fixed beam), subject to the maximum lateral force per unit length p_u at the soil-pipe interface. For this case the axial tension due to arc-length effects is small and neglected. Hence, the maximum strain in the pipe is given by:

$$\varepsilon_b = \pm \frac{p_u W^2}{3\pi E t D^2} \qquad (7.8)$$

Note that M. O'Rourke (1989) assumes that the pipe is fixed at the margins and hence neglects any inward (i.e., axial) movement of pipe at the margin of PGD zone. As a result, Equation 7.7 over-estimates the axial strain in the pipe, as will be shown later.

Liu and M. O'Rourke

Based on the Finite Element results described previously, Liu and M. O'Rourke (1997b) found that pipe strain is an increasing function of ground displacement for ground displacement less than a certain value, δ_{cr}, and pipe strain does not change appreciably thereafter. For example, for $W = 30$ m, as shown in Figure 7.8, the maximum tensile strain is an increasing function of maximum soil displacement up to a value of $\delta = 1.3$ m (4.3 ft). For larger values of δ, the maximum tensile strain remains at a relatively constant value of roughly 0.014. Similar behavior is observed for other widths.

In reality, the pipe resistance to transverse PGD is due to a combination of flexural stiffness and axial stiffness. The analytical relations developed below are for an elastic pipe. Although the inelastic pipe case is more complex, the elastic relations provide a basis for interpreting finite element results and, as will be shown later, are directly applicable to transverse PGD case histories from Niigata.

For small widths of the PGD zone, the critical ground deformation and pipe behavior are controlled by bending. The mechanism is the same as that in the M. O'Rourke (1989) model for the stiff pipe case (i.e., two-end fixed beam with constant distributed load). The critical ground deformation is given by:

$$\delta_{cr-bending} = \frac{5p_uW^4}{384EI} \qquad (7.9)$$

For very large widths of the PGD zone, the pipe behaves like a flexible cable (i.e., negligible flexural stiffness). For this case, the critical displacement is controlled primarily by the axial force. For a parabolic cable shown in Figure 7.24, the relation between the axial force T at the ends and the maximum lateral deformation (or sag) δ is:

$$T = \frac{p_uW^2}{8\delta} \qquad (7.10)$$

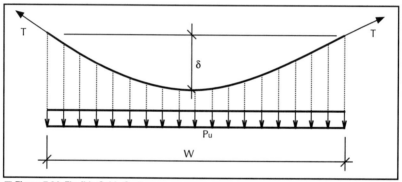

■ Figure 7.24 Flexible Cable System

As shown in Figure 7.10, the ground displacement is larger than the pipe displacement in the middle region of the PGD zone (assumed herein to be $W/2$), over which the maximum transverse resistance per unit length, p_u, at the pipe-soil interface (i.e., the distributed load) is imposed. Taking the "sag" over this middle region to be $\delta/2$, the interrelationship between the tensile force, T, and ground displacement, δ, is given by:

$$T = \pi Dt\sigma = \frac{p_u(W/2)^2}{8(\delta/2)} = \frac{p_uW^2}{16\delta} \qquad (7.11)$$

where σ is the axial stress in the pipe (assumed to be constant within the PGD zone).

Inward movement of the pipe occurs at the margin of the PGD zone due to this axial force. Assuming a constant longitudinal friction force, t_u, beyond the margins, the pipe inward movement at each margin is:

$$\Delta_{inward} = \frac{\pi D t \sigma^2}{2 E t_u} \tag{7.12}$$

The total axial elongation of the pipe within the PGD zone is approximated by the average axial strain given by Equation 7.7 (i.e., arc-length effect) times the width W. This elongation is due to stretching within the zone ($\sigma W/E$) and inward movement at the margins from Equation 7.12. That is,

$$\frac{\pi^2 \delta^2}{4W} = \frac{\sigma W}{E} + 2 \cdot \frac{\pi D t \sigma^2}{2 E t_u} \tag{7.13}$$

The critical ground deformation, $\delta_{cr\text{-}axial}$ for "cable-like" behavior and the corresponding axial pipe stress, σ, can be calculated by simultaneous solution of Equations 7.11 and 7.13. These values are presented in Table 7.1 for three values of the width W and the standard properties mentioned previously (i.e., $D = 0.61$ m, $t = 0.0095$ m, $p_u = 1.0 \times 10^5$ N/m, $t_u = 2.4 \times 10^4$ N/m). Note that the critical ground deformation is controlled by axial force for this case, and that the maximum axial stress at $\delta = \delta_{cr}$ is an increasing function of width of the PGD zone.

■Table 7.1 Critical Ground Displacements and Stresses for "Cable-Like" Elastic Pipe

Item	W = 10 m	W = 30 m	W = 50 m
$\delta_{cr\text{-}axial}$ (Equations 7.11 and 7.13)	0.37 m	1.5 m	2.85 m
σ (Equations 7.11 and 7.13)	92.8 MPa	206 MPa	301 MPa

For any arbitrary width of the PGD zone, somewhat between small and very large, resistance is provided by both flexural (beam) and axial (cable) effects. Considering these elements to be acting in parallel,

$$\delta_{cr} = \cfrac{1}{\cfrac{1}{\delta_{cr-bending}} + \cfrac{1}{\delta_{cr-axial}}} \qquad (7.14)$$

Table 7.2 lists the resulting critical displacements of an elastic pipe ($D = 0.61$ m, $t = 0.0095$ m, etc.) for $W = 10$, 30 and 50 m along with the corresponding elastic finite element results. For $W = 30$ and 50 m, the critical displacement from Equation 7.14 matches that from the elastic finite element model. However, for $W = 10$ m, the critical displacement from Equation 7.14 is an order of magnitude less than that from the elastic finite element model. This is due, in part, to the assumption of a constant transverse load p_u on the pipe for bending effects in the simplified approach. The finite element model, on the other hand, uses transverse elasto-plastic soil springs. As noted previously, one obtains the full load p_u from the soil spring only after 0.06 m (2.4 in) of the relative transverse displacement between the pipe and the soil. Hence, although the fully loaded pipe deflects in bending only 0.015 m (0.6 in) for $W = 10$ m, the bases of the soil springs must move an additional 0.06 m (2.4 in) to obtain the full transverse resistance p_u.

■ Table 7.2 Critical Ground Displacements for Elastic Pipe

Item	W = 10 m	W = 30 m	W = 50 m
$\delta_{cr-bending}$ (Equation 7.9)	0.015 m	1.22 m	9.6 m
$\delta_{cr-axial}$ (Table 7.1)	0.37 m	1.5 m	2.85 m
δ_{cr} (Equation 7.14)	0.015 m	0.67 m	2.2 m
δ_{cr} (F.E. Approach)	0.16 m	0.70 m	2.1 m

Note that the critical displacements for both the simplified elastic and elastic finite element models in Table 7.2 underestimate δ_{cr} for an inelastic pipe shown for example in Figure 7.8. This is due to the fact that for the inelastic pipe model, the steel modulus decreases after yielding, and the pipe must undergo larger deformations such that the strain energy in the pipe equals the work done by the distributed soil springs.

The maximum strains in an elastic pipe are due to the combined effects of axial tension (cable behavior) and flexure (beam behavior), and can be expressed as:

$$\varepsilon_{elastic} = \begin{cases} \dfrac{\pi\delta}{2} \cdot \sqrt{\dfrac{t_u}{AEW}} \pm \dfrac{\pi^2\delta D}{W^2} & \delta \leq \delta_{cr} \\[3ex] \dfrac{\pi\delta_{cr}}{2} \cdot \sqrt{\dfrac{t_u}{AEW}} \pm \dfrac{\pi^2\delta_{cr}D}{W^2} & \delta > \delta_{cr} \end{cases} \qquad (7.15)$$

where A is the pipe cross-sectional area.

7.2.3 COMPARISON AMONG APPROACHES

Table 7.3 presents a summary of the pipe properties and the pipe-soil interaction forces used in the approaches mentioned above.

■ Table 7.3 List of Parameters in Seven Approaches

Item	M. O'Rourke (1989)	Miyajima & Kitaura (1989)	T. O'Rourke (1988)	Suzuki et al. (1988)	Kobayashi et al. (1989)	Liu & O'Rourke (1997b)
Method	Analytical	Analytical	F.E.	F.E.	F.E.	Both
Diameter (m)	0.50, 1.01	0.406	0.61	0.61	0.61	0.61
Thickness	0.0127, 0.0063	0.006	0.0095	0.0127	0.0095	0.0095
Material	Mild steel	Elastic	X-60	X-52	X-42	X-52
t_u (N/m)	-	-	2.4×10^4	1.9×10^4	1.9×10^4	2.4×10^4
P_u (N/m)	8.7×10^4	-	7.7×10^4	1.5×10^5	1.15×10^5	1.0×10^5

Comparing the approaches is difficult since the models have different diameters, wall thickness, pipe-soil interaction parameters, etc. Nevertheless, the bending strains can be compared since the analytical relation for bending strain given in Equation 7.15 suggest that it is only a function of δ, D and W.

Figure 7.25 shows the pipe bending strain for $W > 20$ m, backcalculated from the different approaches, plotted as a function of $\delta D/W$. Herein the bending strain is calculated as one half of the sum of the tensile and compressive pipe strains. Note that the Kobayashi et al. (1989) approach is not included since they did not present compressive strain. In this figure, the straight line with a slope of π^2 is the analytical relation given in Equation 7.6. Note that the Suzuki et al. (1988) as well as the Liu and M. O'Rourke (1997b) results both match the analytical relation fairly well. The T. O'Rourke (1988) results are somewhat less than the analytical results while the Miyajima and Kitaura (1989) results are somewhat higher.

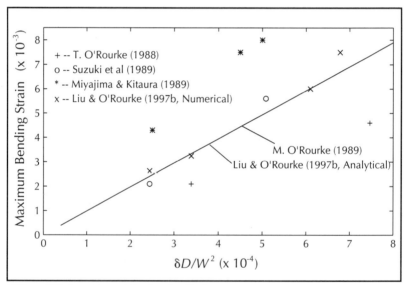

■ Figure 7.25 Comparison of Pipe Bending Strain

Another type of comparison involves the general trends in results from the various approaches. For example, the Liu and M. O'Rourke (1997b) results suggest that axial effects are important in that the tensile strains are larger than the compressive strains. This agrees with the numerical results by T. O'Rourke (1988). In addition, for the three widths considered, the tensile strains are largest for $W = 30$ m which agrees with the numerical results by Suzuki et al. (1988). However, the Liu and M. O'Rourke (1997b) numerical results described above differ from those by T. O'Rourke

(1988), specifically for the width of the PGD zone W = 10 m. Similarly, the Liu and M. O'Rourke (1997b) numerical results differ from those by Suzuki et al. (1988) in that the tensile pipe strains are significantly larger than the compressive strains. It is believed that this difference is due to the comparatively heavy wall thickness used in the Suzuki et al. (1988) model (note as shown in Figure 7.15 that a heavier wall thickness reduces the peak tensile strain but essentially has no effect on the peak compressive strain) in combination with a relatively weak longitudinal soil spring.

7.2.4 COMPARISON WITH CASE HISTORIES

The performance of buried pipelines subject to the transverse PGD during the 1971 San Fernando earthquake provides case histories to test the approaches described above. In the case history shown in Figure 2.11(a), a water transmission pipeline (Pipeline 2) made of Grade-C steel with 1370 mm (54 inch) diameter and 7.9 mm (5/16 inch) wall thickness was subjected to spatially distributed transverse PGD with the maximum ground displacement δ of 0.7 m (2.0 ft.) and a width W of about 400 m (1312 ft.). From Equation 7.6, the bending strain in the pipe with W = 400 m and δ = 0.7 m (2.3 ft) would be 6.0×10^{-5} while the axial tension strain would be 7.6×10^{-6} and the critical displacement δ_{cr} is over 10 m (33 ft). Hence, the maximum tension strain is 6.8×10^{-5} while the net compression strain is 5.2×10^{-5}. Since these values are below the tensile rupture and local buckling strain respectively for the pipe, one expects the pipe would not be damaged by this transverse PGD. The expected behavior matches the observed behavior in that there was no failure within the PGD zone. Note, however, that one break was observed at a location close to but outside the PGD zone, where the pipeline was connected to a ball valve by a mechanical joint at a reinforced concrete vault. According to T. O'Rourke and Tawfik (1983), the mechanical joint was severely deformed and showed signs of repeated impacts. This evidence of repeated impacts suggests that the damage may have been due to wave propagation as opposed to PGD effects.

Although there has been a fair amount of research activity directed at the problem of buried pipe subject to distributed transverse PGD, case histories of continuous pipeline failure due solely to distributed transverse PGD appears fairly unusual.

Although the approximate method (Liu and M. O'Rourke, 1997b) described above is strictly applicable to elastic pipe and widths of 30 m or greater, they prove useful for many realistic design situations. Suzuki and Masuda (1991) present values for the width W and the amount of movement δ, for transverse PGD observed in the Niigata Japan after the 1964 event. Based on roughly 40 separate sites, the amount of ground movement δ ranged from about 0.3 m to 2.0 m, while the width of the PGD zone W ranged from about 100 m to 600 m.

For $W \geq 100$ m (328 ft), steel pipe with $D = 0.61$ m (24 in), $t = 0.0095$ m (3/8 in), the critical ground deformation from Equation 7.14 is 6.0 m (20 ft) or more, which is much larger than the maximum observed ground displacement of 2.0 m (6.6 ft). Also the estimated peak tensile (i.e., combined axial and flexural) strain for $\delta = 2$ m from Equation 7.15 is less than the yield strain for X-grade steel but slightly above the yield strain for GR-B grade steel. Hence, an X-grade pipe behaves elastically and the strain can be estimated by Equation 7.15.

The maximum pipe strains are shown in Figure 7.26 as a function of the ground deformation using both the simplified analytical and numerical models by Liu and M. O'Rourke (1997b). The finite element results for both X-grade and GR-B grade steels are identical for δ less than about 1.6 m. At that location there is a kink in the GR-B curve, indicating the onset of inelastic behavior in that material. Note that the analytical model (i.e., Equation 7.15) results compare favorably with the finite element values.

The approximate analytical approach does overestimate to some degree the peak tensile strain and underestimates the peak compressive strain. This suggests that the estimated axial strains are somewhat too large. However, the differences are relatively small, particularly in light of the accuracy of geotechnical predictions for expected value of the spatial extent and ground movement of PGD zones.

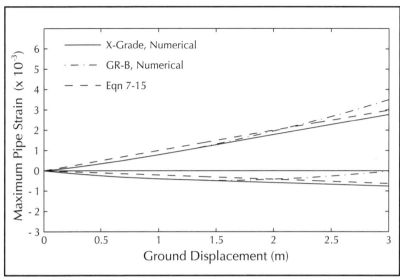

After Liu and M. O'Rourke, 1997b

■ Figure 7.26 Maximum Pipe Strain vs. Ground Displacement for W = 100 m

PIPELINES IN LIQUEFIED SOIL

As mentioned previously, the top of the liquefied soil layer is commonly located below the bottom of the pipe. However, when the pipe is buried in saturated sand such as at a river bed, or in a sea bed, the soil surrounding the pipe may liquefy during strong seismic shaking. In this case, the pipe may deform laterally following the flow of liquefied soil down a gentle slope, or move upward due to buoyancy, especially when a manhole is present or a compressive load acts on the pipe. For example, according to Suzuki et al. (1988) and Takada (1991), a sewage pipe with manhole and a gas pipe (150 mm in diameter) were uplifted out of the ground due to buoyancy in combination with a compressive load caused by longitudinal permanent ground deformation during the 1964 Niigata earthquake. A compressive load can also be induced by temperature change and/or internal operating pressure in a pipe restrained against longitudinal expansion.

When a pipeline is surrounded by liquefied soil, the pipe may move laterally due to the flow of liquefied soil downslope. Using the same model as shown in Figure 7.3, Suzuki et al. (1988) studied the response of a buried pipe, surrounded by liquefied soil, subject to spatially distributed transverse PGD. The presence of the liquefied soil was modeled by assuming that the lateral soil coefficient (K_1) for a pipe surrounded by liquefied soil is some fraction of the corresponding value (K_2) for competent, non-liquefied soil. Figure 7.27 shows the peak pipe strain as a function of the amount of PGD, δ, for three values of the reduction factor. For this plot, the width of the PGD zone is 30 m, while the pipe properties are the same as that listed in Table 7.3 for the Suzuki et al. approach.

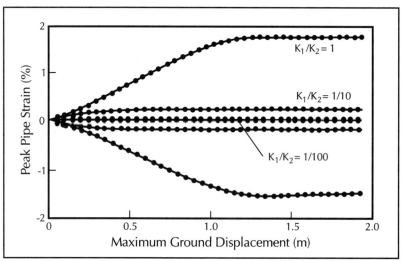

After Suzuki et al., 1988

■ Figure 7.27 Maximum Strain vs. δ for Three Soil Spring Constants

As one might expect, the peak pipe strain for competent non-liquefied soil (i.e., $K_1/K_2 = 1$, also see Figure 7.6 for $W = 30$ m) is in all cases larger than that for liquefied soil. As a rough approximation, the pipe strain for $\delta \geq 1.5$ m (5 ft) is proportional to the soil coefficient reduction factor.

As noted in Section 5.3, the equivalent soil spring coefficient for liquefied soil, according to Takada et al. (1987), ranges from 1/1000 to 1/3000 of that for non-liquefied soil, while other scholars suggest that the ratio is from 1/100 to 5/100. Hence, for the same amount of PGD and width of the PGD zone, a pipe surrounded by liquefied soil is much less likely to be damaged by spatially disturbed transverse PGD. Hence, for design purpose, it seems reasonable to conservatively assume that a pipe subject to spatially distributed transverse PGD is located in a competent non-liquefied soil which overlays the liquefied layer.

7.3.2 VERTICAL MOVEMENT

If the soil immediately surrounding a buried pipe liquefies, the pipe may uplift due to the buoyancy. A few studies have been done regarding this uplifting response. Takada et al. (1987) conducted a series of laboratory tests and estimated the liquefied soil spring constant by combining the test values with analytical solutions. Yeh and Wang (1985) analyzed the dynamic (i.e., seismic shaking and buoyancy effects) pipe response by using a simplified beam-column model for the pipe. They concluded that the dynamic displacement is relatively small (less than 20% of static pipe displacement due to the buoyancy) when the surrounding soil is liquefied.

Using 2 cm (0.8 in) diameter polyethylene pipeline, Cai et al. (1992) carried out a series of laboratory tests and observed pipe response due to soil liquefaction. Figure 7.28 shows the two system models. The model in Figure 7.28 (a) is a pipeline without a manhole while (b) is for a pipeline with a manhole. In both models, the end of the pipe can be either fixed, elastically constrained, or free. The model pipe is 1.2 m (4 ft) in length, which would correspond to a prototype length of 50 m (164 ft) for a prototype diameter of 83 cm (32 in). In these tests, only the shaking and uplifting response can be observed since the simulated ground surface before and after liquefaction is normally flat. That is, lateral response of the pipe is not modeled. For an elastically restrained case, they found that the dynamic strain due to shaking is less than 10% of the static strain due to uplifting and hence can be neglected when estimating the maximum uplifting strain in the pipe. When a manhole and/or an axial compressive force are introduced, the upward response is larger. For elastically constrained ends, the pipe keeps uplifting till a portion of pipe near the center of PGD zone is at the ground surface. However, when a non-liquefied soil layer (60 mm in thickness) is used as cover, the pipe came to rest at the interface of the non-liquefied and liquefied layers.

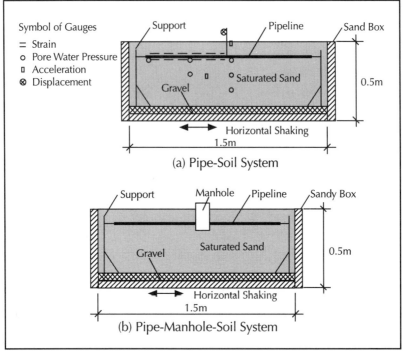

After Cai et al., 1992

■ **Figure 7.28 Model System of Buried Pipeline in Liquefied Soil**

Hou et al. (1990) analyzed the pipe strain due to buoyancy effects by a finite element approach. In the analysis, the non-linearity of both steel material and interaction force at the pipe-soil interface outside the liquefied zone are considered. The uplifting force per unit length, P_{uplift}, acting on the pipe within the liquefied zone, can be expressed as:

$$P_{uplift} = \frac{1}{4}\pi D^2 (\gamma_{soil} - \gamma_{contents}) - \pi Dt\gamma_{pipe} \qquad (7.16)$$

where $\gamma_{soil}, \gamma_{pipe}, \gamma_{contents}$ are the weights per unit volume of liquefied soil, pipe and pipe contents (i.e., water, gas, etc.) respectively. Note that the uplifting force will decrease when a portion of the pipe is at the ground surface.

The pipe is constrained beyond the margins of the liquefied zone by restraint due to the non-liquefied soil. That is, the behavior is similar to a beam, built in at each margin, subject to a uniform upward load. The maximum strain in a steel pipe is shown in Figure 7.29 as a function of the length of the liquefied zone for γ_{soil} = 2.0×10⁴ N/m³,(120 pcf), $\gamma_{contents}$ = 0.8×10⁴ N/m³ (48 pcf), t = 0.0079 m (0.31 in) and three separate pipe diameters. Note that the initial stress ($\sigma_{Initial}$ = 180 MPa) is due to internal operating pressure and/or temperature change.

As shown in Figure 7.29, the maximum pipe strain occurs at a certain width of the liquefied zone, W_{cr}. For the width less than W_{cr}, the pipe strain is an increasing function of the width, while the pipe strain decreases with the increasing width thereafter. The critical width can be estimated by setting Equation 7.8 equal to Equation 7.6 with $\delta = H_c$ (depth from the ground surface to the top of the pipe). That is:

$$W_{cr} = \sqrt[4]{\frac{3\pi^3 EtH_cD^3}{p_u}}$$

(7.17)

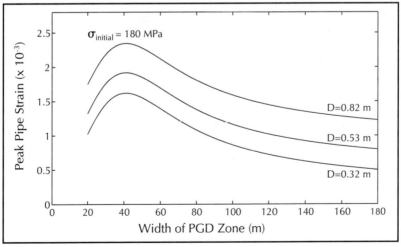

After Hou et al. 1990

■ Figure 7.29 Maximum Strain vs. Width of the Liquefied Zone

For $H_c = 1.2$ m (4 ft), $D = 0.61$ m (24 in) and $p_u = 1.0 \times 10^4$ N/m (57 lbs/in), the estimated critical width of the liquefied zone from Equation 7.17 is 47 m (154 ft), which is slightly larger than the observed critical width of about 42 m (138 ft) in Figure 7.29.

In fact, the buoyancy per unit length given in Equation 7.16 is about 10% of the lateral pipe-soil interaction for a pipe surrounded by non-liquefied soil. That is, it is equivalent to the curve $K_1/K_2 = 1/10$ in Figure 7.27. A comparison of Figures 7.27 and 7.29 indicates that the Suzuki et al.'s results match Hou et al. (1990) reasonably well. That is, for $W \geq 30$ m and $D = 0.61$ m (24 in) in Figure 7.27, the peak pipe strain is about 0.2% while for $D = 0.53$ m (21 in) in Figure 7.29, the peak pipe strain is 0.19%. Since the maximum strain is less than both critical strain of tensile failure and local buckling, the pipe is unlikely to be damaged due to the buoyancy although it may uplift out of the ground when the width of the liquefied zone is large. For situations where a large uplifting displacement is not desirable (for example for submarine pipelines), the following equation derived from the principle of conservation of energy can be used to determine the maximum uplift displacement and/or the spacing for piles or other pipe restraints.

$$\delta_{max}{}^3 + \frac{16I}{A}\delta_{max} - \frac{16p_uW_s^4}{AE\pi^5} = 0 \qquad (7.18)$$

where A is the cross-section area, and W_s is the spacing of the piles as shown in Figure 7.30.

The peak pipe strain is then given by:

$$\varepsilon_{max} = \pm\frac{\pi^2\delta_{max}D}{W_s^2} + \frac{\pi^2\delta_{max}{}^2}{4W_s^2} \qquad (7.19)$$

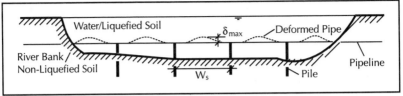

 Water/Liquefied Soil

River Bank
Non-Liquefied Soil

δ_{max} Deformed Pipe

Pipeline

W_s Pile

■ Figure 7.30 Profile of Pipeline Crossing Liquefied Zone

7.4

LOCALIZED ABRUPT PGD

Two patterns of transverse PGD are shown in Figure 7.1. The spatially distributed pattern in Figure 7.1(a) has been discussed extensively above. In relation to the localized abrupt pattern shown in Figure 7.1(b), it was noted that this corresponds more or less to a pair of fault offsets provided that the PGD zone is sufficiently wide. Hence, a key question involves determining the minimum width of the PGD zone, above which the fault crossing models discussed in more detail in Chapter 8 are applicable. Recall that data gathered by Suzuki and Masuda (1991) and shown in Figure 2.8 suggest that PGD zone widths are typically larger than 80 m. Herein a width of 50 m is considered. Figure 7.31 shows the bending moment and axial force in a continuous buried pipeline subjected to a localized abrupt pattern of transverse PGD. The amount of ground movement $\delta = 1.0$ m (3.3 ft) while the width of the PGD zone, W, is 50 m. The pipe and soil properties are $D = 0.61$ m (24 in), $t = 0.0095$ m (3/8 in), $\gamma_{soil} = 1.8 \times 10^4$ N/m³, $\phi = 35°$.

Note that the bending moment is essentially zero over a distance of roughly 20 m (66 ft) near the center of the PGD zone. Hence, in terms of flexure, the continuous pipe behaves as if it was subject to two separate fault offsets, both having a pipe-fault angle of 90°, with no interaction between them. The pipe axial force near the center depends on the width of the PGD zone. It would be zero if the width is large enough such that all the axial force is provided by the friction at the pipe-soil interface within the PGD zone. In this case, the pipe behavior (both tension and flexure) due to a localized abrupt pattern of transverse PGD is the same as that for a pipe crossing a fault with intersection angle of

90°. That is, the procedures described in Chapter 8 could be used directly in this case.

If the axial force near the center of the zone is non-zero (as shown in Figure 7.31(b)), the peak axial force at the margins would be larger than that for two separate fault offsets. That is, by symmetry the center of the zone acts as an effective anchor point, and as will be noted in Chapter 8, an anchor near a fault increases the stress in a pipe subject to movement at the fault.

In summary, there are significant differences in pipe behavior between spatially distributed and localized abrupt transverse PGD. Unfortunately for designers, the authors are not aware of procedures for discriminating, apriori, between these two patterns.

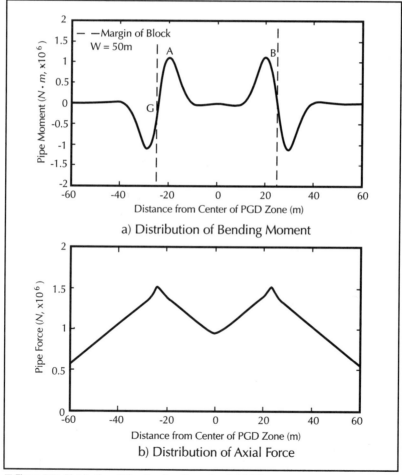

a) Distribution of Bending Moment

b) Distribution of Axial Force

■ Figure 7.31 Distribution of Pipe Bending Moment and Axial Force

RESPONSE OF
CONTINUOUS PIPELINES
TO FAULTING

Chapters 6 and 7 present the response of continuous pipelines subject to longitudinal and transverse PGD, respectively. As mentioned previously, arbitrary PGD can be decomposed into two components, one, parallel to pipe axis (i.e., longitudinal PGD) and the other, perpendicular to pipe axis (i.e., transverse PGD). This chapter presents the response of continuous pipelines subject to fault offsets, which in the general case involves both longitudinal and transverse response.

Two cases are discussed herein. In Case I, pipes are distressed due to bending (caused by transverse component) and axial tensile force (caused by longitudinal component). A normal fault and strike-slip fault with the intersection angle between the fault trace and the pipe axis, β (shown in Figure 8.1), less than 90° are examples of this case. The pipe failure mechanism would be tensile rupture since the fault offset results primarily in tensile strain in the pipe in this case.

In Case II, pipes are distressed due to bending and axial compressive force. A reverse fault and strike-slip fault with the intersection angle, β, between the fault trace and the pipe axis more than 90° are examples of this case. The pipe failure mechanism would be buckling since this type of fault results primarily in compressive strain in the pipe in this case.

Surface faulting has accounted for many pipe breaks during past earthquakes. For example, much of the surface faulting during the 1971 San Fernando earthquake occurred in urban and suburban communities. Although only a half of one percent of the area was influenced by the surface faulting, the fault movements resulted in over 1,400 breaks in water, natural gas and sewer pipelines (McCaffrey and T. O'Rourke, 1983). Among the three fault zones (the Mission Wells segment, the Sylmar segment and the Harding school segment), the 3 km long (1.9 mile) Sylmar segment had the largest ground displacement, which was composed of 1.9 m (6.2 ft) of left-lateral slip, 1.4 m (4.6 ft) of vertical offset and 0.6 m (2.0 ft) of thrust. Most of the left-lateral slip and thrust were concentrated along the southern 25 to 80 m wide (82 to 262 ft) section of the fault zone, where the failure mode of local buckling to the pipe wall was dominant. On the other hand, vertical offsets and extension fractures are predominant in the northern section, where most of the breaks were due to tensile failure.

There are three potential failure modes for a continuous pipeline fault crossing. They are: tensile rupture, local buckling (wrinkling) in compression, and beam buckling in compression. The beam buckling mode is discussed in detail in Section 4.1.3. As noted pipelines are typically buried about 1.0 m (3.0 ft) below the ground surface. Since this burial depth is larger than the critical burial depth shown in Figure 4.5, the pipe wrinkles rather than buckles like a beam when subject to compressive PGD. Hence, this chapter focuses on tensile rupture of a pipe due to bending and tension, and wrinkling of the pipe wall due to bending and compression.

8.1

STRIKE SLIP FAULT

A number of investigations have been performed regarding tensile and bending behavior due to large abrupt fault movements. These include: the Newmark-Hall (1975) approach, the Kennedy et al. approach, and the Wang-Yeh (1985) approach. Herein these analytical approaches are reviewed and the results are compared to those from an FE model, in which both pipe material nonlinearity as well as the nonlinear interaction at the pipe-soil interface are considered.

8.1.1 ANALYTICAL MODELS

Newmark and Hall (1975) apparently were the first to analyze the fault crossing problem. They considered the model shown in Figure 8.1 with a total fault movement δ_f, in which a pipeline intersects a right lateral strike-slip fault at an angle β. For a pipe-fault intersection angle $\beta \leq 90°$, the strike-slip fault results primarily in tensile strain in the pipe. They assume that the pipe is firmly attached to the soil (i.e., no relative displacement between pipe and soil) at two anchor points located at L_a from the fault trace. Anchors correspond to elbows, tie-ins, and other features, which develop substantial resistance to axial movement.

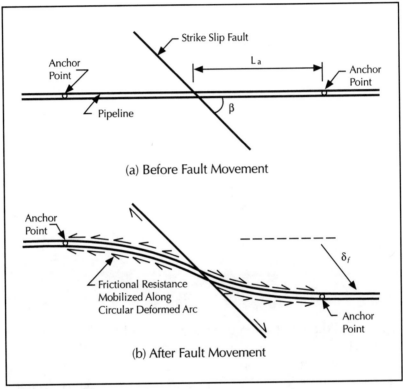

(a) Before Fault Movement

(b) After Fault Movement

After Newmark and Hall, 1975

■ Figure 8.1 Plan View of the Newmark-Hall Model for Pipeline Crossing a Right Lateral Strike-Slip Fault

The authors neglect the bending stiffness of the pipe as well as lateral interactions at the pipe-soil interface. That is, they envision a trench with sloping side walls for which only longitudinal interaction at the pipe-soil interface is considered. The total elongation of the pipe is composed of two components. The first is due to the axial component of fault movement ($\delta_f \cos\beta$). The second is due to arc-length effects caused by lateral component of fault movement ($\delta_f \sin\beta$).

Because of symmetry, only one side of the fault trace is considered. The average pipe strain, $\bar{\varepsilon}$, is:

$$\bar{\varepsilon} = \frac{\delta_f}{2L_a}\cos\beta + \frac{1}{2}\left(\frac{\delta_f}{2L_a}\sin\beta\right)^2 \qquad (8.1)$$

where L_a is the effective unanchored length, that is, the distance between the fault trace and the anchor point.

When no bends, tie-ins or other constraints are located near the fault, the friction forces at the pipe-soil interface provide all the axial resistance. In this case, L_a can be estimated by:

$$L_a = L_e + L_p \qquad (8.2)$$

where L_e is the pipe length over which elastic strain develops, while L_p is the length over which plastic strain develops.

L_e, L_p are given by:

$$L_e = (E_i \varepsilon_y \pi Dt)/t_u \qquad (8.3)$$

$$L_p = [E_p(\varepsilon - \varepsilon_y)\pi Dt]/t_u \qquad (8.4)$$

where ε_y is the yield strain of material, E_i and E_p are the modulus before yield and after yield, ε is the plastic tensile strain in the pipe and t_u is the axial friction force per unit length at the pipe-soil interface. Failure in the Newmark and Hall approach is assumed to occur when the average strain $\bar{\varepsilon}$ is greater than 4%.

For a 0.61 m diameter (24 in) pipe made of X-60 steel, the relation between the tolerable fault movement and the intersection angle using the Newmark and Hall approach is shown in Figure 8.12 along with similar information from other approaches which will be discussed later. The Newmark and Hall model provides valuable insight into the mechanics of this problem, and allows one to evaluate the most influential parameters. However, as will be shown later, this approach over-estimates the tolerable fault movement for pipelines since it uses the average strain as a failure criterion and neglects the lateral interaction at the pipe-soil interface.

Kennedy et al. (1977) extended the ideas of Newmark and Hall, and incorporated some improvements in the method for evaluating the maximum axial strain. They considered the effects of lateral interaction in their analysis. Also, the influence of large axial strains on the pipe's bending stiffness is considered. That is, the pipe bending stiffness becomes very small (roughly 0.5% of the initial stiffness) when axial strain is well beyond the yield strain. As a result, the bending strain in the pipe is relatively small in this approach.

Figure 8.2 presents the Kennedy et al. model, in which the bending strain occurs in the curved region where a constant curvature, $1/R_c$, is assumed.

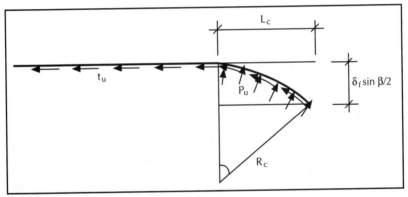

■ **Figure 8.2 Kennedy et al. Model**

The bending strain, ε_b, is expressed as:

$$\varepsilon_b = \frac{D}{2R_c} \qquad (8.5)$$

R_c is the radius of curvature of the curved portion, which can be evaluated by using an analogue to internal pressure in a cylinder:

$$R_c = \frac{\sigma \pi D t}{p_u} \qquad (8.6)$$

where σ is the axial stress at fault crossing, p_u is the lateral soil-pipe interaction force per unit length (see Equation 5.5 for sand or Equation 5.7 for clay).

The total strain in the pipe is given by:

$$\varepsilon = \varepsilon_a + \frac{D}{2R_c} \qquad (8.7)$$

where ε_a is the maximum axial strain due to the elongation of the pipe induced by the fault offset.

The total elongation of the pipe, ΔL, can be estimated by:

$$\Delta L = \delta_f \cos \beta + \frac{(\delta_f \sin \beta)^2}{3L_c} \qquad (8.8)$$

where the first term is the elongation due to the axial component of the fault movement, while the second term is the elongation due to arc-length effects induced by the lateral component of the fault movement, and L_c is the horizontal projection length of the laterally deformed pipe shown in Figure 8.2, which can be approximated by:

$$L_c = \sqrt{R_c \delta_f \sin \beta} \qquad (8.9)$$

Based on the Ramberg Osgood relation in Equation 4.1, the total elongation can be expressed in terms of an integral of axial strain. That is,

$$\Delta L = \frac{2}{E} \int_0^{L_a} \sigma \left[1 + \frac{n}{1+r} \left(\frac{\sigma}{\sigma_y} \right)^r \right] dx \qquad (8.10)$$

Integrating Equation 8.10, one can obtain the relation between the axial movement and effective length L_a. Combining Equation 6.3, Equation 6.4 and Equation 8.8, one can obtain the relation between fault offset and pipe strain for any intersection angle β, which is shown later in Figure 8.12.

Figure 8.3 shows the tolerable fault movement for a 42 in (1.07 m) diameter pipe as a function of unanchored length. The critical tensile strain is 4.5% for Hc = 0.9 m and 3.5% for Hc = 3.0 m due to the substantial increase in bending strains and hoop ovaling for the deeper burial depth. The pipe wall thickness is 0.014 m (0.55 in); it's made of X-60 steel and surrounded by loose to moderately dense sand with $\gamma = 1.76 \times 10^4$ N/m³ (110 pcf) and $\phi = 34°$. The burial depth is 0.91 m (3 ft) from the ground surface to the top of the pipe.

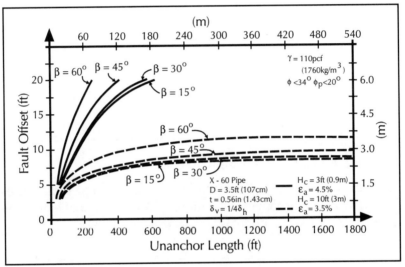

After Kennedy et al. 1977

■ Figure 8.3 Tolerable Fault Movement vs. Unanchored Length

As shown in Figure 8.3, the tolerable fault offset for the pipe is an increasing function of unanchored length and pipe-fault intersection angle, but a decreasing function of burial depth. For the pipe-fault intersection angle of $\beta = 60°$, Figure 8.4 shows the maximum axial strain as a function of unanchored length.

As shown in Figure 8.4, the pipe axial strain is a decreasing function of the wall thickness and unanchored length. For a given wall thickness, the axial strain is an increasing function of the pipe diameter, which is due to the corresponding increase in the interaction forces at the pipe-soil interface. In contrast to the Newmark and Hall approach, the authors consider both the axial and lateral interactions at the pipe-soil interface.

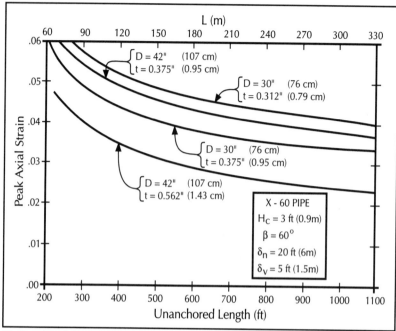

After Kennedy et al. 1977

■ Figure 8.4 Maximum Axial Strain vs. Unanchored Length

Subsequent to the above studies, Wang and Yeh (1985) introduced some additional modifications. The plan view of half of the pipe for Wang-Yeh's model is shown in Figure 8.5.

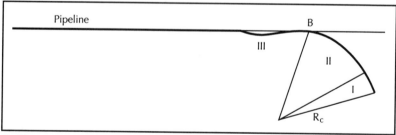

After Wang and Yeh, 1985

■ Figure 8.5 Plan View of Pipeline Crossing Strike-slip Fault

Assuming a constant radius of curvature for the curved portions (i.e., Regions I and II), Wang and Yeh analyze pipe strain based on the equilibrium of forces and deformation compatibility. They also assume that the strains in Regions II and III are elastic, while the strains in Region I are inelastic. For the nominally straight portion of the pipe (Region III), a beam on an elastic foundation theory is used.

For a 1.06 m diameter pipe with 0.014 m in wall thickness, made of X-70 steel and surrounded by loose to moderately dense sand ($\gamma = 1.76 \times 10^4$ N/m^3 (110 pcf), and $\phi = 34°$) with burial depth of 0.91m (3ft) from the ground surface to the top of the pipe, their approach results in the tolerable fault offset of 3.5 m (11.5 ft) for $\beta = 70°$ and 4.6 m (15 ft) for $\beta = 79°$. Note that Wang and Yeh assume that the friction angle at the pipe-soil interface is 20°.

Wang and Yeh apparently neglect the influence of pipe axial stress on pipe bending stiffness, and use the initial modulus to evaluate the bending strain in Regions II and III. As a result, they overestimate pipe bending strain in Region II and conclude that the pipe fails at the Point B as shown in Figure 8.5. This seems counterintuitive since one expects tensile ruptures at or very near the fault crossing location.

8.1.2   FINITE ELEMENT MODELS

Assuming a constant radius of curvature for the curved portion of the pipe, Ariman and Lee (1991) evaluated pipe strain using the finite element method. The pipe is modeled as a thin cylindrical shell which is essentially semi-infinite. For a 42 in diameter pipe made of X-60 steel, they present the bending strain as a function of soil angle of shearing resistance, burial depth, and pipe diameter in Figure 8.6 (a), (b) and (c) respectively. The amount of fault offset is 6.1 m and the intersection angle is 70° in their calculation.

As shown in Figure 8.6, the Ariman and Lee model suggests that the bending strain in the pipe is an increasing function of the soil friction angle, burial depth and pipe diameter.

For a pipe fault crossing at a right angle (i.e., $\beta = 90°$), Meyersohn (1991) evaluates the pipe strain by a finite element program UNIPIP (Tawfik and T. O'Rourke, 1986). Meyersohn's numerical simulations are performed on two pipelines with burial depth of 0.91 m (3 ft) from top of pipe to the ground surface. Pipe 1, with an outside diameter of 1.06 m (42 in) and a wall thickness of 14 mm (0.55 in), and Pipe 2 with a diameter of 0.3 m (12 in) and a wall thickness of 9.5 mm (3/8 in), are unpressurized and made of X-60 steel. Figure 8.7 shows the maximum compressive

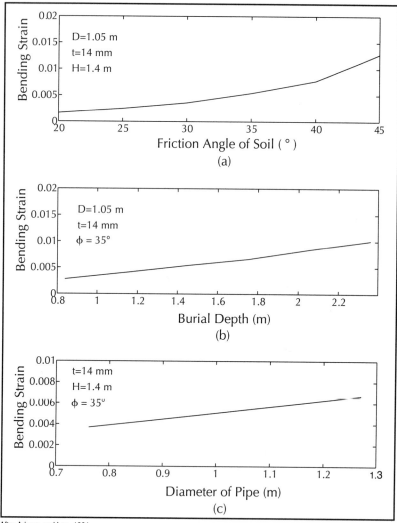

After Ariman and Lee, 1991

■ Figure 8.6 Pipe Bending Strain as a Function of Soil Friction Angle, Burial Depth and Pipe Diameter

strain versus the fault displacement, in which the compressive strain generally increases with an increase in the fault displacement. After the displacement reaches a certain value (for example 2.3 m for Pipe 2), the maximum compressive strain decreases with increase of the displacement since the axial tensile strain becomes larger than yield strain due to the arc-length effects induced by the fault movement, according to Meyersohn (1991).

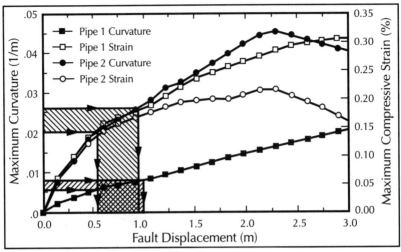

After Meyersohn, 1991

■ **Figure 8.7 Maximum Pipe Compressive Strains as a Function of Fault Offset**

The maximum compressive strain in Figure 8.7 is less than the critical wrinkling strain for the pipes given by Equation 4.3.

In order to evaluate the various approaches, the model shown in Figure 8.8 is used herein to estimate the pipe strain using the ABAQUS finite element program.

As shown in Figure 8.8, the pipeline model is fixed at the point A, which is 500 m (1640 ft) away from the pipe-fault intersection (Point O). This unanchored length is sufficiently long such that both axial strain and bending strain are zero at point A (i.e., the fixed end). Similarly, there is no relative movement between the pipe and the surrounding soil at Point D (i.e., the unrestrained end). All the bases of soil springs to the left of fault trace are fixed. To the right of the fault trace, all the bases of lateral soil springs

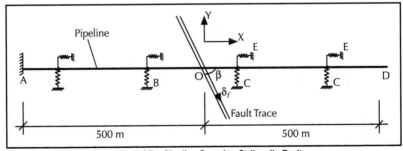

■ **Figure 8.8 Finite Element Model For Pipeline Crossing Strike-slip Fault**

(e.g., Point C) move a distance of $\delta_f \sin\beta$ in the Y direction, while all the bases of axial soil springs (e.g., Point E) move a distance of $\delta_f \cos\beta$ in the X direction.

Considering the non-linear interaction at the pipe-soil interface (Equation 5.1 and Equation 5.5) and the Ramberg Osgood stress-strain relationship (Equation 4.1), the response of an X-60 grade pipe (0.61 m in diameter, 0.0095 m in wall thickness) subject to a strike-slip fault is analyzed.

For a pipe-fault intersection angle of 90°, the distribution of maximum pipe strain is shown in Figure 8.9. The maximum axial strain as well as the peak tensile strain occurs at the intersection where the bending strain is zero (i.e., the point of counterflexure). The bending strain near the intersection (within about 30 m) is relatively constant (i.e., relatively constant curvature as assumed in the Kennedy et al. approach). The peak tensile strain due to the combined effects of axial and bending strains occurs at the intersection.

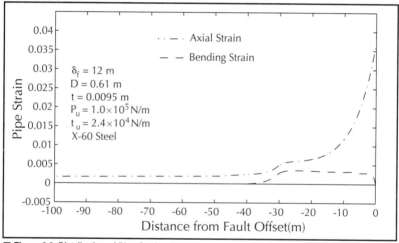

■ Figure 8.9 Distribution of Pipe Strains (for 90° intersection angle)

The peak pipe strain for the same intersection angle (90°) is shown in Figure 8.10 as a function of fault offset.

As shown in Figure 8.10, the peak compression strain occurs for a fault offset of approximately 2 m (6.6 ft) and decreases thereafter. This decrease of compression strain is due to the decrease of bending strain/stiffness caused by the large axial strain. For example, as shown in Figure 8.9 for a fault offset of 12 m (39 ft), the bending strain in the pipe near the fault is a constant value of about 3.6×10^{-3}. Since the axial strain is larger, the net compressive strain is zero.

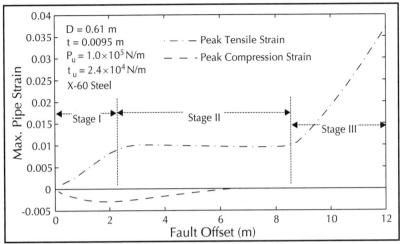

■ Figure 8.10 Pipe Strain vs. Fault Offset (for 90° intersection angle)

Three stages can be identified for the response of a buried pipeline subject to an abrupt lateral fault offset as shown in Figure 8.10. In Stage I (small offsets), both axial and bending strains are important, and both increase with fault offsets. Bending strains are large enough such that there is a non-zero net compressive strain. In Stage II (intermediate offsets), the axial strain is beyond yield, and bending stiffness (and hence bending strain) are decreasing and the net compressive strains approach zero. In Stage III (large offsets), the bending strain remains constant while axial strain increases with increasing fault offsets.

For pipelines made of X-60 and Grade-B steel, the tolerable fault offset is presented in Figure 8.11 as a function of the pipe-fault intersection angle. As shown in Figure 8.11, for an intersection angle less than about 90° (92° for X-60 steel, 90.6° for Grade-B steel), the tolerable fault offset is governed by tensile failure, and increases with the pipe-fault intersection angle. For an intersection angle larger than that angle (Case II), the tolerable fault offset is governed by wrinkling of pipe wall, and the compressive strain reaches the failure criterion at a low value of fault offset. For example, the fault offset of 0.2 m (7.9 in) may cause local buckling to a pipe with 0.61m (24 in) diameter, 0.0095 m (3/8 in) wall thickness and made of Grade-B steel. Note that the solid lines are based on a tensile critical strain of 4%, while the dotted lines are based on a local buckling strain of 0.54% (an average value from Equation 4.3).

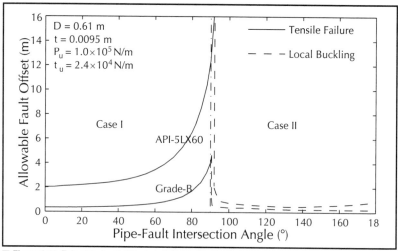

■ Figure 8.11 Tolerable Fault Offset vs. Intersection Angle

8.1.3 COMPARISON AMONG APPROACHES

Figure 8.12 presents the tolerable fault offset from four approaches as a function of the intersection angle between the pipe axis and fault trace.

As shown in Figure 8.12, the results obtained from the ABAQUS numerical approach match that from Kennedy et al.'s analytical approach very well for intersection angles less than 60°. For intersection angles larger than 60°, the tolerable fault offset from the ABAQUS model is somewhat less than that from Kennedy et al.'s approach.

Newmark and Hall's approach overestimates the tolerable fault offset by roughly a factor of 2. This is believed to be due to the use of the average strain for the failure criterion. Note that the maximum strain in the pipe is at least twice the average strain used in their approach. Moreover, they neglect the bending strain in the pipe and the influence of bending strain on the axial stiffness of the pipe. Note that the Newmark and Hall's curve is based on Equation 8.1 with an unanchored length of 50 m (164 ft).

As mentioned previously, for the pipe made of X-70 steel Wang and Yeh's approach leads to tolerable fault offset of 3.5 m (11.5 ft) and 4.6 m (15 ft) for β = 70° and β = 79°, respectively. For the same case, the tolerable fault offset is 15 m (49 ft) for β = 70° and

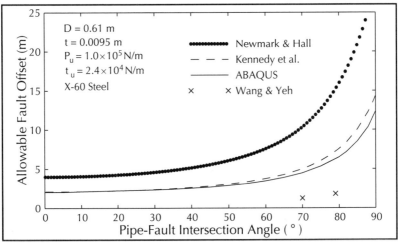

■ Figure 8.12 Comparison of Results from Four Approaches

21 m (69 ft) for $\beta = 79°$ by Kennedy et al.'s approach which, as noted above, appears to match the finite element results. Hence, Wang and Yeh's approach apparently underestimates the tolerable fault offset for the pipeline by a factor of 4. This is believed to be due to their assumption that the pipe strain in Region II (Figure 8.5) is elastic and a reduced bending stiffness at high axial strain is neglected. In fact, the finite element results suggest that the axial strain in that region does exceed the yield strain and the bending stiffness is, in fact, greatly reduced. Hence, Wang and Yeh's approach overestimates the bending strain in the pipe, and underestimates the tolerable fault offset for the pipe.

Results from the Ariman and Lee model as well as those from Meyersohn are presented in terms of bending strain. Since they also used different diameters and wall thicknesses, no comparison is presented herein.

8.1.4   COMPARISON WITH CASE HISTORIES

The 1979 Imperial Valley earthquake provides case histories which can be used to benchmark the finite element approach. During this earthquake, three natural gas pipelines were affected by the localized abrupt offsets at the Imperial fault as shown in Figure 8.13. The maximum co-seismic right lateral slip along the

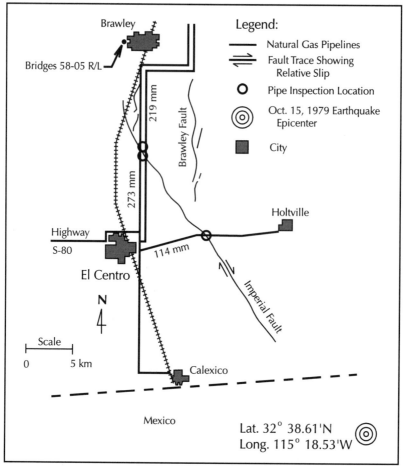

Figure 8.13 Three Gas Pipelines Intersected by the Imperial Fault

fault was 550 to 600 mm (1.8 to 2.0 ft) at Heber Dunes. Up to 290 mm (1.0 ft) of afterslip was measured at McCale Road 160 days after the earthquake, according to Roth et al. (1990).

The material properties of those pipelines, the amount of fault offset as well as the pipe-fault intersection angles are listed in Table 8.1. The No. 56 coating consists of layers of red oxide primer, filled asphalt, two spiral wraps of cellulose acetate, filled asphalt and paper wrapper. The somastic coating is composed of asphalt, aggregate and fiber mixture. In this case study, Equations 5.1 and 5.5 are used for estimating maximum axial and lateral interacting force at the pipe-soil interface. We assume the angle of shear resistance of the sand ϕ = 35° and k = 0.7 for No. 56 coating and k = 0.9 for somastic coating.

Two cases are considered herein. In Case I, the fault is assumed to be a single abrupt fault (i.e., the width of offset zone is zero). In Case II, the 9.6 m (32 ft) of actual fault width is used, and linear distribution of ground movement across the width is assumed. The maximum pipe strains from the finite element model as well as the critical strain are listed in Table 8.1. The critical strain for Holtville-El Centro Line (angle 55°) is taken as a tensile rupture strain of 4%, while the critical strains for Lines 6000 and 6001 are taken as the wrinkling strain from Equation 4.3. The predicted behavior matches the observed behavior in that the maximum strain for Case II (actual width used) are less than the critical values, and the pipes did not, in fact, fail. Note that the tensile strains for the Holtville-El Centro Line for both cases are relatively close. However, for the other two pipelines, the compressive strain for the zero-width fault is much larger that for the 9.6 m-wide fault. This suggests that the width of the fault can be a key parameter, particular for compressional movements. That is, the finite element results suggest that the two pipelines in compression would have wrinkled if the width of the fault were small (e.g. less than about 3.0 m (10 ft)).

■ Table 8.1 Pipe and Fault Properties and Strain Analyses

Characteristic	Holtville-El Centro Line	Line 6000	Line 6001
Diameter	114 mm	219 mm	273 mm
Wall Thickness	5 mm	7 mm	5 mm
Material	A-25 Steel	GR-B Steel	X-42 Steel
Yield Stress	170 MPa	240 MPa	290 MPa
Operating Pressure	2.8 MPa	2.8 MPa	2.8 MPa
Depth of Cover	0.9 m	0.9 m	0.9 m
Weld Type	Acetylene	Electric Arc	Electric Arc
Coating	No. 56	Somastic	No. 56
Fault Offset	0.6 m	0.4 m	0.4 m
Intersection Angle	55°	120°	120°
Max. Tensile Strain (Case I)	0.015	-	-
Max. Tensile Strain (Case II)	0.0126	-	-
Max. Compressive Strain (Case I)	-	> 0.06	0.0335
Max. Compressive Strain (Case II)	-	0.00736	0.00326
Critical Strain	0.04	0.0112	0.0064

Case I: Abrupt offset
Case II: 9.6 m fault width

NORMAL AND REVERSE FAULT

Relatively little analytical work has been done for a pipe across a normal or reverse fault. For a pipe subject to a normal fault, the pipe-soil system is no longer symmetric, and the transverse inter-action force at the pipe-soil interface for downward movement of the pipeline is much larger than that for upward movement, based on Equation 5.9 to Equation 5.16. Kennedy et al.'s approach can still be used to estimate the pipe strain. In this case, an average lateral interaction force for upward and downward movements can be used to estimate the bending strain from Equations 8.5 and 8.6, and the axial strain is the same as before.

However, for a pipe subject to a reverse fault, it appears that no analytical approach is currently available. The ASCE TCLEE Committee on Gas and Liquid Fuel Lifeline (1984) suggests using the finite element method.

The behavior in such cases is difficult to generalize, in part because there are two angles of intersection (the angle in plan between the fault and the pipeline, as well as the dip angle of the fault) as well as the aforementioned asymmetric nature of the soil resistance in the vertical plane.

RESPONSE OF
SEGMENTED
PIPELINES TO PGD

In this chapter, the response of segmented pipelines subject to PGD will be discussed. Segmented pipes typically have bell and spigot joints and can be made of cast iron, ductile iron, steel, concrete or asbestos cement. As indicated in Section 4.2, there are three main failure modes for segmented pipelines: axial pull-out at joints, crushing of the bell and spigot joints, and round flexural cracks in the pipe segment away from the joints.

Similar to the response of continuous pipelines, the behavior of a given buried segmented pipeline is a function of the type of PGD (e.g. longitudinal or transverse), the amount of ground movement δ, the spatial extent of the PGD zone and the pattern of ground movement within the zone.

In reference to the type, Suzuki (1988) concluded that damage due to longitudinal PGD was more common than damage due to transverse PGD based on the observed damage to segmented gas pipelines during the 1964 Niigata earthquake. In these cases, the joints were pulled out in the tension region and buckled in the compression region.

In terms of the pattern, if the ground movement within the PGD zone is relatively uniform (i.e., an idealized block pattern of longitudinal PGD in Figure 6.1(a)), one expects that a few pipe joints near the head and toe of the zone would have to accommodate essentially all the abrupt differential ground movement. On the other hand, if the ground movement varies within the PGD zone (i.e., an idealized ridge pattern of longitudinal PGD in Figure 6.1(c)), the rate of change along the segmented pipeline leads to an "equivalent" ground strain. One expects that all joints within the zone, to a greater or lesser extent, would then experience relative axial displacement.

In this chapter, the response of segmented pipe to longitudinal and transverse PGD as well as fault offsets are discussed.

L O N G I T U D I N A L   P G D

As with continuous pipeline, longitudinal PGD induces axial effects in segmented pipeline, specifically axial strain in the pipe segments and relative axial displacement at the joints. However, in contrast to the response of continuous pipelines, damage to segmented pipelines subject to longitudinal PGD typically occurs at pipe joints since the strength of the joints is generally less than the strength of the pipe itself. Whether the joints fail depends on the strength and deformation capacity of the joints as well as the characteristics of the PGD.

One particularly important characteristic is the pattern of longitudinal PGD. Herein, two types of patterns are considered in detail. For the distributed deformation case (such as the idealized ridge pattern in Figure 6.1(c)), ground strain exists over a significant portion of the PGD zone. For the abrupt deformation case (such as the idealized block pattern in Figure 6.1(a)), relative movement exists only at the margins of the PGD zone, and the ground strain between the margins is zero.

9 . 1 . 1   D I S T R I B U T E D   D E F O R M A T I O N

The response of segmented pipelines subject to a distributed deformation pattern of longitudinal PGD is similar to that for segmented pipelines subject to wave propagation in that the spatially distributed PGD results in a region of ground strain. That is, the ridge, asymmetric ridge and ramp patterns in Figure 6.1 result in ground strain over the whole length of the PGD zone, while the Ramp-Block pattern results in uniform ground strain over a portion (i.e., length βL) of the zone. For example, the ground strain for the ridge pattern is:

$$\varepsilon_g = \frac{2\delta}{L} \qquad (9.1)$$

By assuming that pipe segments are rigid and all of the longitudinal PGD is accommodated by the extension or contraction of the joints, the average relative displacement at the joints is given by the ground strain times the pipe segment length, L_o.

$$\Delta u_{avg} = \frac{2\delta L_o}{L} \tag{9.2}$$

Although Equation 9.2 represents the average behavior, the joint displacements for uniform ground strain varied somewhat from joint to joint due to variation in joint stiffness. That is, a relatively flexible joint is expected to experience larger joint displacements than adjacent stiffer joints. Using realistic variations of joint stiffness, El Hmadi and M. O'Rourke (1989) determined, as presented in Table 9.1, the mean joint displacement, Δx, in centimeters, and coefficients of variation, μ, in percentage, as a function of ground strain for various diameters of Cast Iron pipe with lead caulked joints (CI) and Ductile Iron pipe with rubber gasketed joints (DI). The values in Table 9.1 assume that the pipe segment length L_o for all types was 6.0 m (20 ft).

■ Table 9.1 Mean Joint Displacement and Coefficient of Variation for Segmented Pipe Subject to Uniform Ground Strain

Ground Strain	CI D=40 cm (16 in)		CI D=76 cm (30 in)		CI D=122 cm (48 in)		DI D=40-122 cm (16-48 in)	
	$\Delta \bar{x}$ (cm)	μ(%)	$\Delta \bar{x}$ (cm)	μ(%)	$\Delta \bar{x}$ (cm)	μ(%)	$\Delta \bar{x}$ (cm)	μ(%)
0.001 (1/1000)	.54	64	.56	54	.58	52	.59	2
0.002 (1/500)	1.14	56	1.17	49	1.17	43	1.19	2
0.005 (1/200)	2.92	39	2.95	24	2.97	14	3.00	1
0.007 (1/150)	4.12	26	4.16	19	4.16	16	4.19	1

Note that the mean values for both CI and DI pipes are about equal to the value given in Equation 9.2 (that is $\varepsilon_g L_o$).

As shown in Table 9.1, μ for DI joints is quite small in comparison to that for CI joints. This is due to the fact that DI joints are substantially more flexible than CI joints. As a result, the joint opening for DI pipelines would be relatively constant over the length of the PGD zone.

To gauge the effects of a distributed deformation pattern of longitudinal PGD on segmented pipe, expected joint openings are calculated. M. O'Rourke et al. (1995) present a summary of longitudinal PGD pattern observed in Noshiro City after the 1983 Nihonkai Chubu event. The minimum ground strain due to the distributed longitudinal PGD is 0.008. The corresponding joint opening is 5 cm (2 in), which is larger than the joint capacity of typical segmented pipelines as noted in Section 4.2 (i.e., segmented joints typically leak for relative displacement on the order of half the total joint depth). Hence, typical segmented pipelines are vulnerable and consideration should be given to replacement by continuous pipelines or segmented pipelines with special joints (having large contract/expansion capacity and/or anti-pull-out restraints) when crossing a potential longitudinal PGD zone.

The potential for damage due to something other than joint pull-out of simple bell and spigot joints is more difficult to evaluate. For example, tensile failure of various types of restrained joints or crushing of simple bell and spigot joints typically involves some slippage in the joint before significant load is transferred across the joint. In this regard, the expected behavior of concrete pipe joints in compression is discussed in Chapter 11.

9.1.2   ABRUPT DEFORMATION

As used herein, abrupt longitudinal PGD refers to ground movements with large relative offsets at localized points. The block pattern in Figure 6.1(a) is an example. In this case, the ground strain is zero away from the margins of the PGD zone, there is a tensile opening or gap at the head of the zone and a localized compressive mound at the toe. The ramp and ramp-block patterns in Figure 6.1 (b) and (d) also have an abrupt offset, but for these patterns at only one end of the PGD zone.

At the head of the zone (i.e., the tension gap), pipeline failure for typical bell and spigot joints is probable. In the simplest model, one expects joint leakage or pull-out if the relative joint displacement corresponding to leakage or pull-out respectively is less than the ground offset (that is δ in Figure 6.1).

For the 17 idealized block, ramp or ramp-block patterns studied by M. O'Rourke et al. (1995), the abrupt offset, δ, was 1.2 m (4 ft) or larger. Hence, one expects joint pull-out in typical segmented pipe at tension gaps at least for the examples of longitudinal PGD considered by M. O'Rourke et al. (1995).

An axially restrained pipe at a tension gap or a restrained or unrestrained pipe at a compression mound behaves like a continuous pipe subject to longitudinal PGD as discussed in Chapter 6. That is, in the simplest model, the flexibility at the joint itself (i.e., due to stretching of bolts for a restrained joint in tension, or joint push-in for a joint in compression) is neglected. Failure is possible in the pipe segment or at the joint closest to the tension gap or compression mound. For the "no-joint-flexibility" assumption, the pipe segment behaves like a continuous pipe. Potential failure modes and strains are discussed in detail in Chapter 6. The axial force at the joint closest to the abrupt offset is the smaller of $t_u L/2$ or $t_u L_e$ as shown in Figures 6.5 and 6.6.

TRANSVERSE PGD

In considering the response of segmented pipelines subject to transverse PGD, one must differentiate between spatially distributed transverse PGD and localized abrupt transverse PGD as sketched in Figure 7.1. Localized abrupt PGD is a special case of fault offset (intersection angle of 90°).

9.2.1 SPATIALLY DISTRIBUTED PGD

For segmented pipelines subject to spatially distributed transverse PGD, the failure modes include round cracks in the pipe segments and crushing of bell and spigot joints due to the bending, and pull-out at the joint due to axial elongation (i.e., arc-length effects).

For an assumed sinusoidal variation of ground movement across the width of the PGD zone as given by Equation 7.3 and shown in Figure 7.23, M. O'Rourke and Nordberg (1991) studied the maximum joint opening due to both joint rotation and axial extension of segmented pipelines. Figure 9.1(a), (b) present a pipeline subject to transverse PGD, where Δx_t and $\Delta \theta$ are the joint extension and relative joint rotation between the adjacent segments.

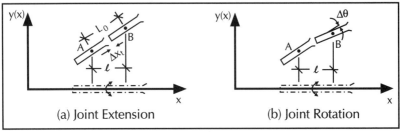

(a) Joint Extension   (b) Joint Rotation

After M. O'Rourke and Nordberg, 1991

■ **Figure 9.1 Plan View of Segmented Pipeline Subject to Distributed Transverse PGD**

Assuming that the pipe segments are rigid (i.e., EA=∞, EI= ∞) and that the lateral displacement at the midpoint of the rigid pipe segment exactly matches the spatially distributed PGD at that point, they developed the relative axial displacement at a joint.

$$\Delta x_t = \frac{L_o}{2} \left(\frac{\pi \delta}{W} \sin \frac{2\pi x}{W} \right)^2 \tag{9.3}$$

where x is the distance from the margin of the PGD zone and L_o is the pipe segment length.

The axial displacements are largest for joints near $x = W/4$ and $3W/4$. Hence, a pure joint-pull-out failure mode is most likely at the locations $W/4$ away from the center of the PGD zone. The peak axial displacement is given by:

$$\Delta x_t = \frac{L_o}{2} \left(\frac{\pi \delta}{W} \right)^2 \tag{9.4}$$

Assuming that the slope of the rigid pipe segment exactly matches the ground slope at the segment midpoint, the joint opening due to the joint rotation, Δx_r, is as follows:

$$\Delta x_r = \begin{cases} \dfrac{\pi^2 \delta D L_o}{W^2} \cos \dfrac{2\pi x}{W} & \Delta x_t > \Delta \theta \cdot D / 2 \\ \dfrac{2\pi^2 \delta D L_o}{W^2} \cos \dfrac{2\pi x}{W} & \Delta x_t < \Delta \theta \cdot D / 2 \end{cases} \tag{9.5}$$

where D is the pipe diameter.

This function is a maximum at $x = 0$, $W/2$ and W. Hence, a pure joint rotation failure and/or flexural round cracks are more likely at the margins and middle point of the PGD zone.

The total maximum opening at one side of a joint, Δx, due to transverse PGD, is simply the sum of axial extension plus rotation effects. However, the axial and rotational components are largest at different points as discussed previously. Combining these effects, the resulting maximum joint opening is:

$$\Delta x = \begin{cases} \dfrac{\pi^2 L_o \delta^2}{W^2}\left[\dfrac{2D}{\delta}\right] & 0.268 \leq D / \delta < 3.73 \\[4mm] \dfrac{\pi^2 L_o \delta^2}{2W^2}\left[1 + (D / \delta)^2\right] & \text{Others} \end{cases}$$

(9.6)

This relation for the maximum joint opening is plotted in Figure 9.2. Note that the maximum joint opening is an increasing function of both the δ/W ratio and the D/δ ratio.

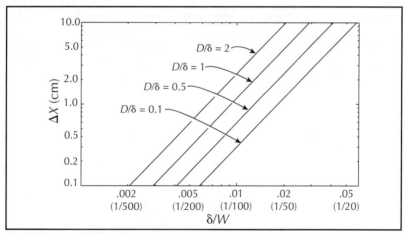

After M. O'Rourke and Nordberg, 1991

■ Figure 9.2 Maximum Joint Opening for Segmented Pipe Subject to Distributed Transverse PGD

Consider the observed spatially distributed transverse PGD during the 1964 Niigata and 1983 Nihonkai Chubu events (Suzuki and Masuda, 1991) shown in Figure 2.8. Observed values for the δ/W ratio range from 0.001 to 0.01 with 0.003 being a typical value. The amount of ground movement δ ranged from 0.2 to 2.0 m (0.66 to 6.6 ft) with 1.0 m (3.3 ft) being a typical value. Hence, from Figure 9.2, the corresponding maximum joint opening (i.e., using upper bound values of δ/W=0.01 and δ=2.0 m) would be 2.5 cm (1 in) or less for pipe diameter of 4.0 m (157 in) (i.e., D/δ=2) or less. Hence, significant leakage or pull-out failure at these joints due to the axial movement is unlikely to occur for segmented pipelines subject to spatially transverse PGD shown in Figure 2.8. However, for segmented pipelines with rigid joints, some leakage at joints is likely to occur since leakage may occur at joint opening of 2.0 mm (0.08 in) based on the laboratory tests (Prior, 1935).

9.2.2   FAULT OFFSETS

Both experimental and analytical results are available for segmented pipelines subject to fault offset (i.e., local abrupt differential ground movement transverse to the pipe axis). For example, Takada (1984) performed a laboratory test to analyze the response of segmented pipelines subject to transverse PGD. Figure 9.3 presents a sketch of the sinking soil box (dimension 10 m×1 m×1.5 m), in which a 169 mm (nominal 6 in) diameter Ductile Iron pipeline is surrounded by loose sand. The vertical offset is produced by decreasing the height of the six jacks which support the movable box. Two cases were studied in their tests. In Case A, the pipeline is composed of three longer segments, while in Case B it is composed of five shorter segments.

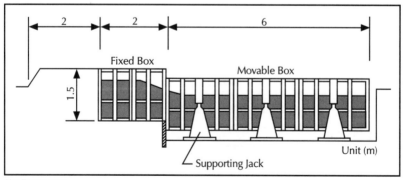

After Takada, 1984

■ Figure 9.3 Model Box for Segmented Pipeline Subject to Transverse PGD

Figure 9.4 shows the maximum pipe stress, which occurs directly over the offset, versus the ground subsidence for both cases.

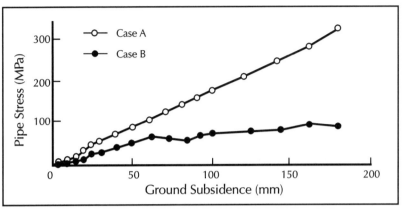

After Takada, 1984

■ Figure 9.4 Maximum Pipe Stress vs. Ground Subsidence for Case A and Case B

As shown in Figure 9.4, the stresses in the pipeline with smaller-length segments (Case B) are much less than those for large-length segments (Case A) particularly for large values of the offset.

For the geometry studied by Takada, that is, a pipe at 90° with respect to the fault or offset plane, flexural stresses in the pipe dominate. If one assumes rotationally flexible joints (i.e., no moment transfer across the joint), the portion of the pipe segment on one side of the fault plane acts as a cantilever beam subject to a distributed loading along its length due to transverse pipe-soil interaction forces and a concentrated load at its end (i.e., at the joint) due to shear transfer across the joint. Smaller pipe stress in Case B are due, in part, to the shorter cantilever length.

Analytical results for segmented pipes are also available. T. O'Rourke and Trautmann (1981) developed a simplified analytical method for evaluating the response of segmented pipelines subject to fault offset. They assume that segments are rigid and joints accommodate the ground deformation. Figure 9.5 shows the plan view of a segmented pipeline subject to fault offset.

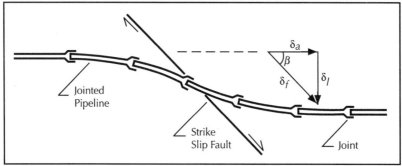

After T. O'Rourke and Trautmann, 1981

■ Figure 9.5 Plan View of Segmented Pipeline Subject to Fault Offset

The tolerable fault displacement can be obtained by:

$$\delta_f = min\begin{cases}\delta_a \sec\beta \\ \delta_l \csc\beta\end{cases} \qquad (9.7)$$

where δ_a is the pull-out capacity of the joint (axial deformation) near the fault offset, δ_l is the lateral deformation capacity, which depends on the joint rotation ability and is calculated by finite element simulations for typical ductile and cast iron pipelines.

The optimal orientation of the pipeline, $\beta_{optimal}$, can be defined by:

$$\beta_{optimal} = \frac{\delta_l}{\delta_a} \qquad (9.8)$$

A pipeline would fail by joint pull-out when the intersection angle between the pipe axis and the fault trace is less than $\beta_{optimal}$, while the pipeline would fail by bending when the intersection angle is larger than $\beta_{optimal}$.

T. O'Rourke and Trautmann (1981) plotted the tolerable fault offset for segmented pipelines as a function of the intersection angle as shown in Figure 9.6. Similar to the response of continuous pipelines subject to fault offset, the tolerable fault offset for pipelines with either restrained or unrestrained joints is an increasing function of β for the intersection angle less than the optimal value, and decreases thereafter. For example, the optimal intersection angle for pipe with mechanical joints is about 70°. According to T.

O'Rourke and Trautmann, the decrease in capacity for $\beta > \beta_{optimal}$ is caused by the larger bending moments developed in the pipeline for large intersection angles.

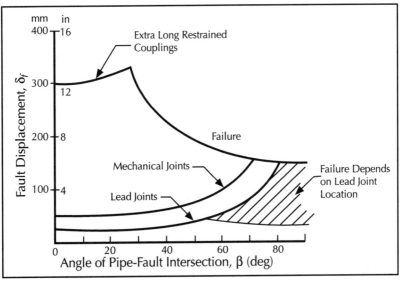

After T. O'Rourke and Trautmann, 1981

■ Figure 9.6 Tolerable Fault Offset vs. Intersection Angle

Note that pipe with extra long restrained coupling are particularly effective only when the intersection angle is small. At these small intersection angles, axial effects dominant and the expansion capability of the special joints is useful. However, at large intersection angles ($\beta > 60°$), where flexural effects govern, the capacity of mechanical and special joints is similar.

Post earthquake observation of segmented pipe response to fault offsets would be useful for case history verification of available analytical procedures.

RESPONSE OF BURIED CONTINUOUS PIPELINES TO WAVE PROPAGATION

There have been some events, such as the 1964 Puget Sound, 1969 Santa Rosa, 1983 Coalinga and 1985 Michoacan earthquakes, for which seismic wave propagation was the predominate hazard to buried pipelines. For example, the damage ratio for the water supply system in the Lake Zone (soft soil zone) of Metropolitan Mexico City of about 0.45 repairs/km has been attributed to wave propagation effects in the 1985 Michoacan event.

As discussed in Chapter 3, when a seismic wave travels along the ground surface, any two points located along the propagation path will undergo out-of-phase motions. Those motions induce both axial and bending strains in a buried pipeline due to interaction at the pipe-soil interface. For segmented pipelines, damage usually occurs at the pipe joints. Although seismic wave propagation damage to continuous pipelines is less common, the observed failure mechanism is typically local buckling.

This and the following chapter focus on buried pipe response due to wave propagation effects. The existing methods for evaluating the response of continuous pipelines as well as the behavior at elbows and tees are discussed and compared in this chapter. The following chapter discusses similar issues for segmented pipelines.

10.1 STRAIGHT CONTINUOUS PIPELINES

In general, the axial strain induced in a straight continuous pipeline depends on the ground strain, the wavelength of the travelling waves and the interaction forces at the pipe-soil interface. For small to moderate ground motion, one may simply assume that pipe strain is equal to ground strain. However, for large ground motion, slippage typically occurs at the pipe-soil interface, resulting in pipe strain somewhat less than the ground strain.

10.1.1 NEWMARK APPROACH

Simplified procedures for assessing pipe response due to wave propagation were first developed by Newmark (1967), and have since been used and/or extended by a number of authors (e.g., Yeh, 1974). Newmark's approach is based on three assumptions. The first assumption, which is common to most all the deterministic approaches, deals with the earthquake excitation. The ground motion (that is, the acceleration, velocity and displacement time histories) at two points along the propagation path are assumed to differ only by a time lag. That is, the excitation is modeled as a traveling wave. The second assumption is that pipeline inertia terms are small and may be neglected (Wang and M. O'Rourke, 1978). Experimental evidence from Japan (Kubo, 1974) as well as analytical studies (Sakurai and Takahashi, 1969, Shinozuka and Koike, 1979) indicate that this is a reasonable engineering approximation. The third assumption is that there is no relative movement at the pipe-soil interface and hence, the pipe strain equals the ground strain.

Figure 10.1 shows a pipeline subject to S-wave propagation in a vertical plane having an angle of incidence γ_s with respect to the vertical.

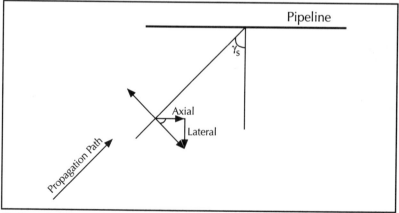

After Meyersohn,1991

■ Figure 10.1 Pipeline Subject to S-wave Propagation

For this case, the ground strain parallel to the pipe axis is:

$$\varepsilon_g = \frac{V_m}{C_s} \sin \gamma_s \cos \gamma_s \tag{10.1}$$

where V_m is the peak ground velocity and C_s is the shear wave velocity.

In terms of Equation 3.8, $V_m \cos \gamma_s$ is the ground velocity parallel to the pipe axis and, as noted in Equation 3.5, $C_s/\sin \gamma_s$ is the apparent propagation velocity with respect to the ground surface and the pipeline axis.

Similarly, for R-wave, the ground strain parallel to the pipe axis is:

$$\varepsilon_g = \frac{V_m}{C_{ph}} \tag{10.2}$$

Since bending strain in a pipe due to wave propagation is typically a second order effect, our attention is restricted to axial strain in the pipe. Equations 10.1 and 10.2 overestimate pipe strain, especially when the ground strain is large. For those cases, slippage occurs at the pipe-soil interface and the pipe strain is less than the ground strain.

10.1.2   SAKURAI AND TAKAHASHI APPROACH

In relation to Newmark's assumption regarding pipeline inertia, Sakurai and Takahashi (1969) developed a simple analytical model for a straight pipeline surrounded by an infinite elastic medium (soil). They used D'Alembert's principle to handle the inertia force. For a pipeline subject to the ground displacement u_g, the equilibrium for the pipe segment is:

$$\rho \frac{\partial^2 u_p}{\partial t^2} - E \frac{\partial^2 u_p}{\partial z^2} = K_g(u_g - u_p) \tag{10.3}$$

where u_p is the displacement of the pipeline in z direction (longitudinal direction), assumed to be the direction of wave propagation,

K_g is the linear soil stiffness per unit length as shown in Figure 5.2 and ρ is the mass density of pipe material.

The analytical results from Equation 10.3, which do not consider slippage at the pipe-soil interface, indicate that the pipe strain is about equal to free field strain and hence, the inertia effects are negligible. This result regarding inertia terms is not surprising in light of the fact that the unit weight of a fluid filled pipe is not greatly different from that of the surrounding soil.

10.1.3 SHINOZUKA AND KOIKE APPROACH

In relation to Newmark's assumption regarding no relative displacement at the pipe-soil interface, Shinozuka and Koike (1979) modify Equation 10.3 as follows:

$$\rho \frac{\partial^2 u_p}{\partial t^2} - E \frac{\partial^2 u_p}{\partial z^2} = \tau_s \, / \, t \qquad (10.4)$$

where τ_s is the shear force at the pipe-soil interface per unit length and t is the pipe wall thickness.

Neglecting the effects of inertia, Shinozuka and Koike (1979) developed a conversion factor between ground and pipe strains. For the case of no slippage at the pipe-soil interface (i.e., the soil springs remain elastic), the conversion factor is,

$$\beta_0 = \frac{1}{1 + \left(\dfrac{2\pi}{\lambda} \right)^2 \cdot \dfrac{AE}{K_g}} \qquad (10.5)$$

That is, the pipe strain is β_o times the ground strain. This result holds as long as the shear strain at the pipe-soil interface, $\gamma_{o'}$

$$\gamma_0 = \frac{2\pi}{\lambda} \frac{Et}{G} \varepsilon_g \beta_0 \qquad (10.6)$$

is less than the critical shear strain, γ_{cr}, beyond which slippage occurs at the pipe-soil interface. The critical shear strain as estimated by Shinozuka and Koike is:

$$\gamma_{cr} = \frac{t_u}{\pi DG} \qquad (10.7)$$

In their analysis, Shinozuka and Koike (1979) assumed that the critical shear strain is 1.0×10^{-3}. That is, for $\gamma_0 \leq 1\times10^{-3}$, slippage will not take place, while for $\gamma_0>1\times10^{-3}$, slippage occurs at the pipe-soil interface.

For large amounts of ground movement, i.e., $\gamma_0>\gamma_{cr}$, the ground to pipe conversion factor is:

$$\beta_c = \frac{\gamma_{cr}}{\gamma_0} q\beta_o \qquad (10.8)$$

where q is a factor which range from 1 to $\pi/2$ and quantifies the degree of slippage at the pipe-soil interface. That is for slippage over the whole pipe length $q = \pi/2$.

The pipe axial strain is then simply calculated by:

$$\varepsilon_p = \beta_c \cdot \varepsilon_g \qquad (10.9)$$

10.1.4   M. O'ROURKE AND EL HMADI APPROACH

Also in relation to Newmark's "no relative displacement assumption", M. O'Rourke and El Hmadi (1988) use a somewhat different approach to estimate the maximum axial strain induced in a continuous pipe due to wave propagation.

Consider a model of a buried pipeline shown in Figure 10.2. The pipe has cross-sectional area A and modulus of elasticity E. The soil's resistance to axial movement of the pipe is modeled by a linear spring with stiffness K_g and a slider which limits the soil spring force to the maximum frictional resistance t_u at the pipe-soil interface. If the system remains elastic, that is the pipe strain remains below its yield strain and the soil spring force remains below t_u, the differential equation for the pipe axial displacement $U_p(x)$ is:

$$\frac{d^2}{dx^2} U_p(x) - \beta^2 U_p(x) = -\beta^2 U_g(x) \qquad (10.10)$$

where $\beta^2 = K_g/(AE)$ and $U_g(x)$ is the ground displacement parallel to the pipe axis.

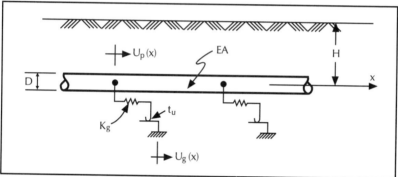

After M. O'Rourke and El Hmadi, 1988

■ Figure 10.2 Continuous Pipeline Model

If the ground strain between two points separated by a distance L_s is modeled by a sinusoidal wave with wavelength $\lambda = 4L_s$, the ground deformation $U_g(x)$ (i.e., displacement of the base of the soil springs) is given by:

$$U_g(x) = \varepsilon_g L_s \sin \frac{\pi x}{2L_s} \qquad (10.11)$$

where ε_g is the average ground strain over a separation distance L_s. The pipe strain is then given by:

$$\varepsilon_p = \frac{dU_p}{dx} = \frac{\pi}{2} \varepsilon_g \frac{\beta^2}{\beta^2 + \left(\dfrac{\pi}{2L_s}\right)^2} \cos \frac{\pi x}{2L_s} \qquad (10.12)$$

The elastic solution given in Equation 10.12 holds as long as the pipe strain is below its yield strain and the maximum force in the soil spring is less than the frictional resistance at the pipe-soil interface. That is,

$$\varepsilon_g L_s \left[1 - \frac{\beta^2}{\beta^2 + \left(\dfrac{\pi}{2L_s} \right)^2} \right] < \frac{t_u}{K_g} \tag{10.13}$$

From Equation 10.13, a slip strain ε_s is defined as

$$\varepsilon_s = \frac{t_u}{K_g L_s} \left[\frac{\beta^2 + \left(\dfrac{\pi}{2L_s} \right)^2}{\left(\dfrac{\pi}{2L_s} \right)^2} \right] \tag{10.14}$$

For moderately dense backfill, the slip strain is plotted in Figure 10.3 as a function of separation distance L_s. In this plot, two different nominal diameters of X-60 grade pipe, D = 30 cm (12in) and 91 cm (36 in), as well as two different burial depths, H = 0.75 m (2.5 ft) and 1.5 m (5 ft), are considered.

Since the slippage strains are less than the strains which would results in pipe damage, propagation damage to continuous pipe typically involve some slippage at the pipe-soil interface.

With this in mind, M. O'Rourke and El Hmadi consider the upper bound case where slippage occurs over the whole pipe length. For a wave with wavelength λ, the points of zero ground strain (points A and B), as shown in Figure 10.4, are separated by a horizontal distance of $\lambda/2$. Assuming a uniform frictional force per unit length t_u, the maximum pipe strain at point C due to friction is given by:

$$\varepsilon_p = \frac{t_u L_s}{AE} \tag{10.15}$$

where $L_s = \dfrac{\lambda}{4}$.

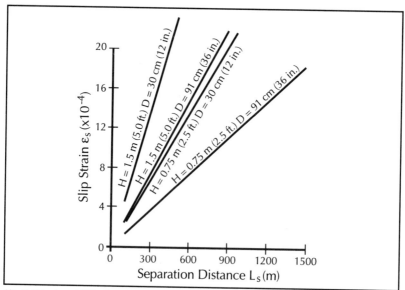

After M. O'Rourke and El Hmadi, 1988

■ Figure 10.3 Slip Strain vs. Separation Distance for Moderately Dense Sand Backfill

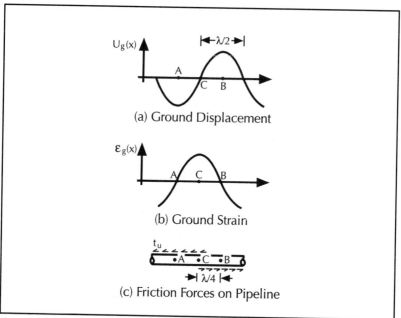

M. O'Rourke and El Hmadi, 1988

■ Figure 10.4 Friction Strain Model for Wave Propagation Effects on Buried Pipelines

For R-waves, M. O'Rourke and El Hmadi developed an analysis procedure to estimate the maximum pipe strain. This procedure compares axial strain in the soil to the strain in a continuous pipeline due to soil friction along its length. It is assumed that the soil strain is due to R-waves propagating parallel to the pipe axis. Due to the dispersive nature of R-wave propagation (i.e., phase velocity an increasing function of wavelength), the soil strain is a decreasing function of separation distance or wavelength. The pipe strain due to the friction at the pipe-soil interface is an increasing function of separation distance or wavelength. At a particular separation distance (that is, for a particular wavelength), the friction strain matches the soil strain. This unique strain then becomes the peak strain which could be induced in a continuous pipeline by R-wave propagation. Figure 10.5 shows both the ground strain and the pipe strain as function of the separation distance for an elastic pipe.

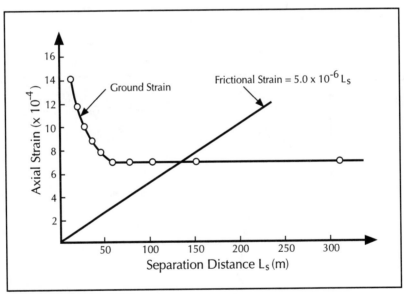

■ Figure 10.5 Frictional Strain and Ground Strain vs. Separation Distance

As shown in Figure 10.5, for shorter quarter-wavelength separation distances, the pipeline frictional force acts over the whole length (i.e., from A to B in Figure 10.4) and hence, the pipe strain is linearly proportional to the quarter-wavelength separation distance. However, at longer quarter-wavelength separation distances, the pipe frictional force acting only near Points A and B results in a pipe strain equal to the ground strain at Point C. Note that this procedure for R-waves conservatively assumes that the peak ground velocity, V_{max}, applies to all frequencies (wavelengths) of R-wave propagation and that all frequencies (wavelengths) are present in the record.

10.1.5   COMPARISON   AMONG   APPROACHES

A comparison of the three approaches for a continuous pipe subject to wave propagation are presented in this subsection. The comparison is based on R-wave propagation having a dispersion curve with $\upsilon = 0.48$ shown in Figure 10.6. The peak particle velocity is taken as 0.35 m/s. The ground strains at three frequencies, from Equation 10.2, are presented in Table 10.1 along with the estimated strain in a straight pipeline with $D=1.07$ m (42 in) and $t = 8$ mm (5/16 in).

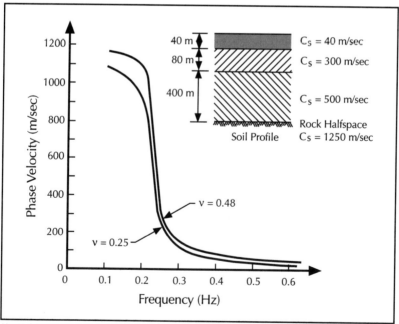

After M. O'Rourke and Ayala, 1990

■ Figure 10.6 Dispersion Curve and Ground Profile for the 1985 Mexico Earthquake Case History

f (Hz)	C_{ph} (m/s)	Wavelength (m)	ε_g ($\times 10^{-3}$)	Pipe Strain ($\times 10^{-3}$)			
				Newmark	Shinozuka & Koike q=1	q=π/2	O'Rourke & El Hmadi
0.2	900	4500 m	0.39	0.39	0.31	0.39	0.39
0.3	137	456 m	2.5	2.5	0.77	1.3	1.3
0.4	92	230 m	3.8	3.8	0.4	0.6	0.6

As shown in Table 10.1, three approaches result in essentially the same pipe strain when the ground strain is small. In this case the pipe and soil move together and pipe strain is equal to ground strain since no slippage occurs. However, for large ground strains, the pipe strains from the Shinozuka and Koike approach as well as the M. O'Rourke and El Hmadi approach are both much less than ground strain. That is, although the ground strains are larger, the quarter wavelength distances over which the soil friction forces act are comparatively small. Note that Shinozuka and Koike's approach for full slippage case with $q = \pi/2$ is essentially the same as M. O'Rourke and El Hmadi's. For $q = 1$ in the Shinozuka and Koike approach, slippage occurs only over a portion of the pipe, and the corresponding pipe strains are lower bounds.

10.1.6 COMPARISON WITH CASE HISTORIES

During the 1985 Michoacan earthquake, a welded steel pipeline with $D = 107$ cm (42 in), $t = 0.8$ cm (5/16 in) and made of API 120 X-42 steel was damaged at several locations within the Lake Zone in Mexico City. As a case study, M. O'Rourke and Ayala (1990) estimated the compressive stress in the pipe due to R-wave propagation.

Figure 10.6 shows the dispersion curve for the fundamental R-wave, corresponding to the subsoil conditions of the Lake Zone in the Mexico City (M. O'Rourke and Ayala, 1990). Note that the generalized ground profile for this site consists roughly of a 40 m layer of soft clay with a shear wave velocity of 40 m/s. Under this layer, there are two stiffer strata with shear wave velocities of 300 and 500 m/s respectively. At the bottom is rock with a shear wave velocity of 1250 m/s.

For a pipe surrounded by loose sand with γ=110 lb/ft³ (17.2 kN/m³) and a coefficient of friction μ=0.5, the estimated compressive strain using M. O'Rourke and El Hmadi's procedure was about 0.002. The corresponding plot of the ground strain and friction strain is shown in Figure 10.7. Note in this figure, the friction strain is proportional to the quarter wavelength (i.e., separation distance) for strains less than about 0.001 (λ/4≈100 m). For larger separation distances, although the axial force is still proportional to separation distance, the strain is not since we are now in the non-linear portion of the stress-strain diagram for the steel. The local buckling strain is estimated to be about 0.0026 based upon D/t=134. That is, the analytical procedure suggests that the pipeline was very close to buckling. Note that the pipeline did, in fact, suffer a local buckling failure at several locations separated by distances of 300 to 500 m (984 to 1640 ft). This corresponds reasonably well with the 130 m (426 ft) quarter wavelength distance in Figure 10.7. That is, high compression regions are a wavelength apart, or 520 m (1706 ft) for the critical quarter-wavelength of 130 m (426 ft).

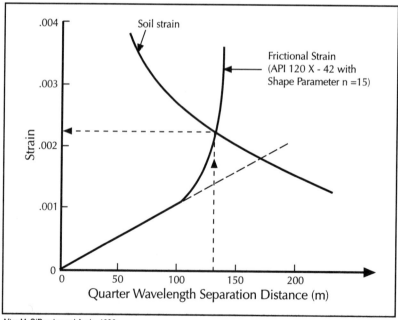

After M. O'Rourke and Ayala, 1990

■ Figure 10.7 Soil and Friction Strain for Ciudad Nezahualcoyotl Pipeline

BENDS AND TEES

A pipe network is typically composed of straight pipeline sections, and interconnecting bends, tees and crosses. The presence of these elements can produce additional bending strains at these interconnects and possibly lead to pipe damage. This section will focus on the effects of bends and tees.

10.2.1 SHAH AND CHU APPROACH

Considering the interaction forces at the pipe-soil interface, Shah and Chu (1974), as well as Goodling (1983), developed analytical formulae for forces and moment at elbows and tees. Figure 10.8 shows the forces acting on a pipeline and pipe deformation near the bend. The traveling wave is assumed to be propagating parallel to Element 1 with ground motion also parallel to Element 1 (e.g., R-waves). Element 2 is modeled as a beam on an elastic foundation with lateral soil stiffness K_g.

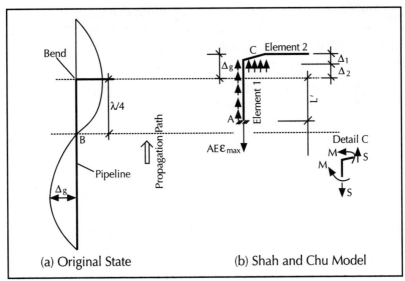

(a) Original State   (b) Shah and Chu Model

After Shah and Chu, 1974

■ Figure 10.8 Displacement and Forces in a Pipe with an Elbow

Shah and Chu (1974) assumed that the pipe and ground strains are equal at a location (Point A in Figure 10.8 b) where no relative displacement occurs at the pipe-soil interface. Denoting the distance from this location to the bend as L' shown in Figure 10.8, Shah and Chu (1974) as well as Goodling (1983) then estimated the maximum axial force in Element 1 at the bend (shear force in Element 2) by:

$$S = \varepsilon_{max} AE - t_u L' \qquad (10.16)$$

The moment and flexural displacement at a bend can be then calculated as:

$$M = \frac{S}{3\zeta} \qquad (10.17)$$

$$\Delta_1 = \frac{4\zeta S}{3K_g} \qquad (10.18)$$

where $\zeta = \sqrt[4]{K_g / (4EI)}$ and L' is the effective slippage length at the bend.

The effective slippage length, L', can be calculated based on displacement compatibility at the bend. That is, within the distance L', total ground deformation (taken as $\varepsilon_{max} L'$) is accommodated by the lateral displacement of Element 2, Δ_1, and axial deformation of Element 1, $\frac{SL'}{AE} + \frac{t_u L'^2}{2AE}$. For a long leg case (i.e., long Element 1), this compatibility condition yields,

$$L' = \frac{4AE\zeta}{3K_g} \left(\sqrt{1 + \frac{3\varepsilon_{max} K_g}{2t_u \zeta}} - 1 \right) \qquad (10.19)$$

Similarly, Figure 10.9 shows the forces and deformation for a tee, again for a wave propagating path parallel to Element 1.

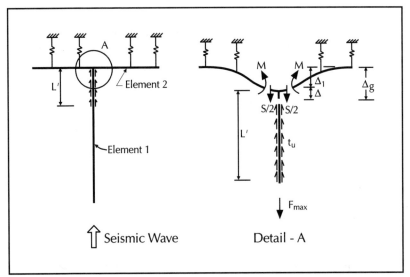

After Goodling, 1983

■ Figure 10.9 Representation of Forces, Moments and Displacement at a Tee

Using the same procedure as that for bends, Shah and Chu estimate the force, moment and displacement for a tee by the following equations:

$$S = \varepsilon_{max} AE - t_u L' \qquad (10.20)$$

$$M = \frac{S}{2\zeta} \qquad (10.21)$$

$$\Delta_1 = \frac{\zeta S}{K_g} \qquad (10.22)$$

$$L' = \frac{AE\zeta}{2K_g} \left(\sqrt{1 + \frac{4\varepsilon_{max} K_g}{t_u \zeta}} - 1 \right) \qquad (10.23)$$

Note that Shah and Chu (1974) as well as Goodling (1983) assume pipe strain is equal to the maximum ground strain at Point A (Figure 10.8). Based upon the previous discussion of straight pipe response to wave propagation, this assumption is likely only true for small ground strains. Furthermore, they estimate the total ground displacement simply by the maximum ground strain times the effective length L'. This implies that the ground strain is constant over the length L' which only applies for a wavelength many times larger than the length L'.

10.2.2 SHINOZUKA AND KOIKE APPROACH

Assuming a pipe moving with the soil at the location with zero ground movement (Point B in Figure 10.10), Shinozuka and Koike (1979) developed simple equations to estimate pipe strain at bends based on structural analysis similar to that discussed above. In their analysis, the effective length, that is, the L' term, is assumed to be a quarter wavelength, and the forces are obtained, as in the previous model, by displacement compatibility at the bend. The axial force, S, can be then expressed by:

$$S = \frac{3\lambda}{8\pi} \cdot \frac{K_g}{\zeta} \cdot \frac{1}{1+Q}\left(1 - \beta_c\right)\varepsilon_g \qquad (10.24)$$

where $Q = \dfrac{3}{16} \dfrac{K_g \lambda}{AE\zeta}$.

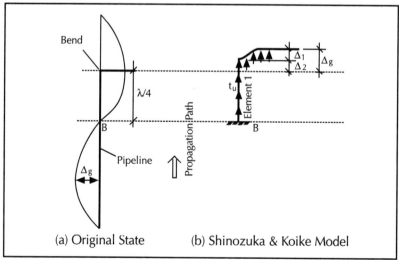

(a) Original State   (b) Shinozuka & Koike Model

After Shinozuka & Koike, 1979

■ Figure 10.10 Displacement and Forces in a Pipe at an Elbow

The moment and displacement at the bend can then be calculated by Equations 10.17 and 10.18. Note that the total ground deformation within the quarter wavelength is calculated by integrating the pipe strain. That is,

$$\Delta_g = \frac{\lambda}{2\pi} \cdot \varepsilon_{max}$$ (10.25)

Similarly, the axial force S in Element 1 for a tee is:

$$S = \frac{\lambda}{2\pi} \cdot \frac{K_g}{\zeta} \cdot \frac{1}{1 + \frac{4}{3}Q}\left(1 - \beta_c\right)\varepsilon_g$$ (10.26)

1 0 . 2 . 3   F I N I T E   E L E M E N T   A P P R O A C H

In order to independently evaluate the assumptions which underlay the existing approaches, the finite element model shown in Figure 10.11 was used. In this numerical model, axial and lateral soil springs are used to model the interaction at the pipe-soil interface. Element 1 is 600 m long and hence, considered appropriate for wavelength of roughly 600 m or less. The quasi-static seismic excitation is modeled by displacing the bases of the soil

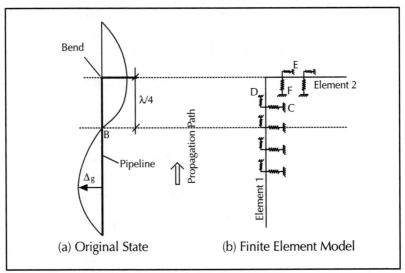

(a) Original State   (b) Finite Element Model

■ Figure 10.11 FE Model for Elbow Subject to Wave Propagation

springs. For example, point F moves Δ_g in the direction of wave propagation, while point E does not move. For Element 1, the movement of the bases (e.g., point D) of the longitudinal soil springs varies along the pipe matching the sinusoidal pattern as shown in Figure 10.11(a).

A steel pipe with diameter $D = 0.76$ m (30 in), wall thickness $t = 0.0095$ m (3/8 in) is considered. The assumed seismic excitation is an R-wave with $V_{max} = 0.36$ m/s propagating parallel to Element 1.

10.2.4   COMPARISON   AMONG APPROACHES

Results from the finite element approach described in Section 10.2.3 are compared to the existing analytical approaches in this section. For an elbow, the force, moment and displacement at the elbow due to travelling wave effects are listed in Table 10.2 using the Shah and Chu approach, the Shinozuka and Koike approach as well as the finite element approach described above. Note that two cases are considered in Table 10.2. In this first case (Case I) the ground strain and wavelength are taken as 0.29×10^{-3} and 244 m respectively. While in Case II, $\varepsilon_g = 1.8 \times 10^{-3}$ and $\lambda = 100$ m.

From Equation 10.19, the effective length for the large ground strain, small wavelength case is 233.3 m by the Gooding/Shah and Chu approach. Since this effective length is much larger than a quarter wave length, that approach can not be used. Note that the effective length by Shinozuka and Koike matches relatively well with the finite element results for both cases considered here.

As shown in Table 10.2, for a small ground strain case, the peak pipe strain at the elbow by Shah and Chu is larger than that by both Shinozuka and Koike and the finite element method. This is due to the fact that Shah and Chu overestimate the ground deformation, and simply assume the maximum pipe strain equal to maximum ground strain. On the other hand, Shinozuka and Koike's approach underestimates the pipe strain at the elbow. This is due to the fact that the axial soil stiffness they suggested ($K_s = 2\pi G = 2\pi\rho C_s^2 = 4.1 \times 10^9$ N/m² (595 kips/in²)) is much larger than that ($K_g = t_u/x_u = 8.3 \times 10^6$ N/m² (1.2 kips/in²)) in the finite element model. For example, by using $K_g = 8.3 \times 10^6$ N/m² (1.2 kips/in²) in Shinozuka and Koike's approach, for the small ground strain case, the peak strain at the elbow is estimated to be 4.9×10^{-5}, which matches the numerical strain (4.5×10^{-5}) very well.

Overall, the comparison in Table 10.2 suggests that, of the available analytical approaches, the Shinozuka and Koike method appears more appropriate.

■ Table 10.2 Comparison For Bend

Approach	Case	λ (m)	ε_g ($\times 10^{-3}$)	Effect Length (m)	Δ_g (cm)	Δ (cm)	S (N)	M (N · m)	Peak Strain
Goodling, Shah & Chu	II	244	0.29	42.8	1.24	0.61	4.2×10^4	5.9×10^4	1.4×10^{-4}
	I	100	1.8	233.3	-	-	-	-	-
Shinozuka & Koike	II	244	0.29	$61\,(\frac{\lambda}{4})$	1.1	.001	71.8	100.6	0
	I	100	1.8	$25\,(\frac{\lambda}{4})$	2.9	2.4	1.64×10^5	2.3×10^5	5.5×10^{-4}
Finite Element	II	244	0.29	60	1.1	0.3	3.2×10^4	3.7×10^4	4.5×10^{-5}
	I	100	1.8	23.6	2.9	2.4	1.9×10^5	2.9×10^5	5.9×10^{-4}

RESPONSE OF SEGMENTED PIPELINES TO WAVE PROPAGATION

As noted previously, seismic wave propagation has caused damage to segmented pipelines. Damage most frequently occurs at joints, tees and elbows. The corresponding failure modes include pull-out at joints, crushing of bell-spigot joints as well as circumferential cracks due to bending. In this chapter, analytical approaches for estimating both the axial and bending strain in straight pipelines are reviewed. Observed expansion/contraction behavior of joints at elbows and connections is also presented. Finally, a somewhat special case of the influence of liquefied soil on the dynamic response of segmented pipelines is discussed.

11.1 STRAIGHT PIPELINES / TENSION

For a long straight run of segmented pipe, the ground strain is accommodated by a combination of pipe strain and relative axial displacement (expansion/contraction) at pipe joints. As noted by Iwamoto et al. (1984), since the overall axial stiffness for segments is typically much larger than that for the joints, the ground strain results primarily in relative displacement of the joints. As a first approximation, assuming that the pipe segment axial strain can be neglected (i.e., rigid segment) and that all joints experience the same movement, the maximum joint movement Δu is:

$$\Delta u = \varepsilon_{max} \cdot L_0 \tag{11.1}$$

where L_0 is the pipe segment length and ε_{max} is the maximum ground strain parallel to the pipe axis, given, for example, by Equation 3.8.

For ground motion perpendicular to the pipe axis, the maximum relative rotation at pipe joints can be estimated by:

$$\Delta\theta = \kappa_g \cdot L_0 \qquad (11.2)$$

where κ_g is the maximum ground curvature given, for example, by Equation 3.9. Equation 11.2 assumes that the bending strain in pipe segments is small and that all joints experience the same relative rotation.

In relation to the rigid segment assumption, Wang (1979) determined the joint deformation and pipe strain using an analytical model shown in Figure 11.1, in which the joint is modeled as a linear spring with axial stiffness K_g.

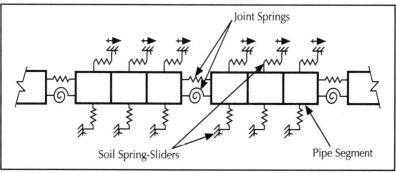

After Wang, 1979

■ Figure 11.1 Model of Segmented Pipeline

Figures 11.2 and 11.3 present the joint opening and maximum axial strain at pipe segments respectively, as a function of joint stiffness for the East West component of the 1940 El Centro event. The assumed pipe diameter is 45.7 cm (18 in), the pipe segment length is 6.1 m (20 ft), the axial soil spring has stiffness of 23.4 MPa (3400 lbs/in^2) and the propagation velocity is taken as 244 m/sec (800 ft/sec). As one expects, the joint opening is a decreasing function of the joint stiffness while the pipe strain is an increasing function of joint stiffness. That is, for a small joint stiffness, the ground deformation is accommodated primarily by joint opening.

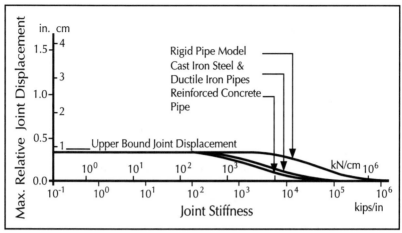

After Wang, 1979

■ **Figure 11.2 Joint Opening due to Wave Propagation vs. Joint Stiffness**

The peak ground velocity for the record used by Wang is 0.37 m/sec, and the value given by Equation 11.1 is close to the upper bound joint opening of 0.85 cm (0.32 in) in Figure 11.2.

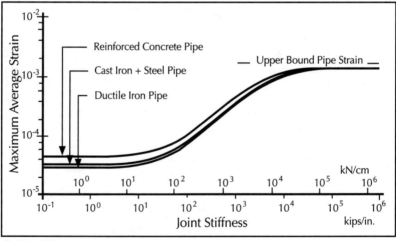

After Wang, 1979

■ **Figure 11.3 Maximum Axial Strain at Pipe Segments vs. Joint Stiffness**

The Wang model correctly captures the trend of decreasing joint opening with increasing joint stiffness. However, it assumes an equivalent linear joint stiffness while laboratory tests suggest that joint axial behavior is non-linear. Furthermore, for a given stiffness, the relative displacement at each joint in the model is the same. That is, it does not capture the variation in displacement from joint to joint. This variation from joint to joint is considered important since even for relatively large amount of wave propagation damage, only a few joints require repair. For example, from Figure 1.3, one expects roughly 0.9 wave propagation repair per km for a peak particle velocity of 50 cm/sec. This suggests one repair for every 182 joints if the pipe segment length is 6.1 m (20 ft). That is, since it is reasonable to assume some variation in response from joint to joint, the few joints with largest response control damage as opposed to joints with "average" response. With this in mind, El Hmadi and M. O'Rourke (1990) considered a model somewhat similar to that in Figure 11.1, in which the joint properties vary from joint to joint and the soil properties vary from pipe segment to pipe segment. Specifically, a cast iron pipe with lead

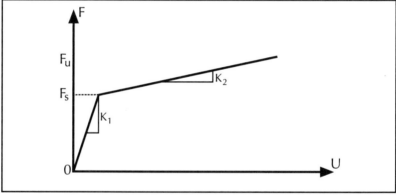

After El Hmadi and M. O'Rourke, 1990

■ Figure 11.4 Axial Force-Displacement Curve for a Lead Caulked Joint

caulked joints subject to tensile ground strain was considered. The assumed force-deformation relation for the joint in tension is shown in Figure 11.4. The expected variation in the joint slippage force, F_s, was based upon results by T. O'Rourke and Trautmann (1980).

A quasi-static approximation to the seismic wave propagation environment is modeled by displacing the base of the soil spring sliders in the longitudinal direction. A simplified Monte Carlo simulation technique is used to establish the characteristics of force-displacement relationships at each joint and soil restraint along each pipe segment. Figure 11.5 shows the joint deformation as a function of ground strain for a segmented pipe with a diameter of 0.41 m (16 in).

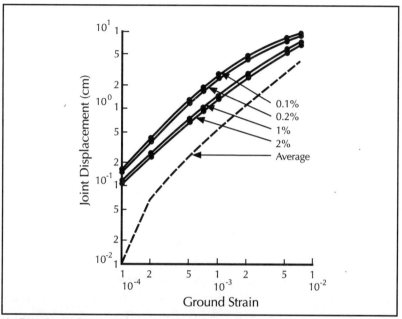

After El Hmadi and M. O'Rourke, 1990

■ Figure 11.5 Relative Joint Displacement vs. Ground Strain for 41 cm Diameter Cast Iron Pipe with Lead Caulked Joints

As shown in Figure 11.5, the average joint displacement is approximately equal to the product of the ground strain times the pipe segment length. However, for the data in Figure 11.5, one in a hundred joints (1% probability of exceedence) have joint displacement about three times the average value while for the 0.1% exceedence probability (one in a thousand), the joint opening is about five times the average. This information, coupled with the probability of leakage as a function of the normalized joint opening, as shown in Figure 4.11, allows one to establish an analytically derived estimation of joint pull-out damage (repair per kilometer) as a function of ground strain. Note however, that this approach requires information, typically derived from laboratory tests, on the expected variability of joint properties.

El Hmadi and M. O'Rourke found that the variability of joint displacement was a decreasing function of pipe diameter. That is, at larger diameters, the joint displacements with 1% and 0.1% probability of exceedence were closer to the average value. This suggests that the damage ratio (repair/km) for joint pull-out in a cast iron pipe with lead caulked joints is a decreasing function of pipe diameter.

STRAIGHT PIPELINES/ COMPRESSION

Extensive damage to concrete pipelines has occurred when these elements are subject to compressive ground strain. For wave propagation resulting in compressive ground strain, the failure mode of interest is crushing (i.e., telescoping) at pipe joints. For concrete pipelines, a strength of materials model, based upon the pipe wall thickness, diameter and concrete strength, was used to establish the joint crushing force as discussed in Section 4.2.2. Figure 11.6 presents the force-displacement relationship for pipe joints subject to compressive load. This relation is based upon a series of laboratory tests on reinforced concrete cylinder pipelines (RCC) with rubber gasketed joints by Bouabid (1995).

These tests indicate that the joint behaves in a sigmoidal fashion before "lock up" (at about 0.3 inches as shown in Figure 11.6). The joint compressive displacement, Δu_{ult}, at lock-up typically ranges from 0.125 to 0.375 in (0.32-0.95 cm) with corresponding loads of 3.5 to 4.5 kips (16-20 kN).

When subject to compressive ground strain ε_g, the response of a segmented pipe is complicated by the presence of joints. Significant axial force can be transferred from joint to joint only if the contraction of the joint is Δu_{ult} (i.e., the joint is fully closed). If there are n fully closed joints in sequence and the ground strain is assumed uniform over the corresponding number of pipe segments, the pipe segment compressive strain is:

$$\varepsilon_p = \varepsilon_g - \frac{n}{n+1}\frac{\Delta u_{ult}}{L_o}$$

$$(11.3)$$

where L_o is the length of pipe segment.

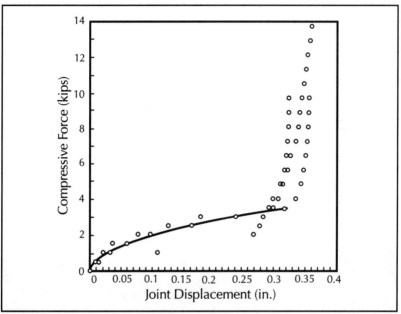

■ Figure 11.6 Force-Displacement Relationship for Reinforced Concrete Cylinder Pipe Joints

The upper bound value corresponds to $n=1$ and

$$\varepsilon_p = \varepsilon_g - \frac{1}{2}\frac{\Delta u_{ult}}{L_o}$$

$$(11.4)$$

The lower bound value corresponds to $n=\infty$ and

$$\varepsilon_p = \varepsilon_g - \frac{\Delta u_{ult}}{L_o}$$

$$(11.5)$$

For non-liquefied soil, the half-wavelength of seismic waves (corresponding to $(n+1)L_o$) is generally larger than 120 m, hence, the pipe strain is expected to be closer the lower bound value (i.e., Equation 11.5).

Using the Monte Carlo technique, M. O'Rourke and Bouabid (1996) developed fragility relations shown in Figure 11.7 for three types of concrete pipe subject to axial compression. These three pipes are 30 inch-diameter reinforced concrete cylinder pipe (30" RCC), 48 inch-diameter prestressed lined cylinder pipe (48" LCP) and 60 inch-diameter prestressed embedded cylinder pipe (60" ECP).

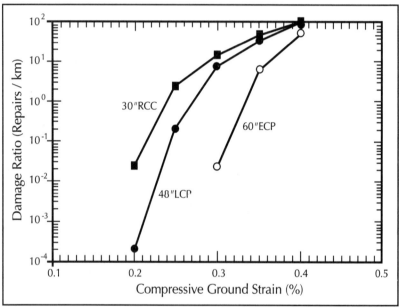

■ Figure 11.7 Analytical Fragility Curves for Concrete Pipe Subject to Compressive Ground Strain

In this case, the variation in joint crushing thresholds was based upon the cross-sectional area near the joint and an assumed normal distribution of concrete strength (mean strength of 5 ksi (34.5 MPa) and 7% coefficient of variation). During the 1985 Michoacan Earthquake, the average ground strain in the portions of Mexico City with a significant inventory of large diameter concrete pipe was estimated by M. O'Rourke and Bouabid to be about 0.0025. Note from Figure 11.7, the expected damage ratio for the 30" RCC (76 cm) pipe is about 2.4 repairs/km, but only 0.22 repairs/km for the 48" LCP (1.22 m) pipe. The estimated value for the 48" pipe, in fact, matches relatively well with the observed damage ratio of 0.20. Also most of the pipelines in this Mexico City comparison are 48 inches in diameter.

ELBOWS AND CONNECTIONS

There appear to have been relatively little analytical research on the wave propagation behavior of bends and elbows in segment pipe systems. However, measurements by Iwamoto et al. (1985) suggest that joint openings at bends and elbows are, in fact, different from those in long straight runs of pipe. For example, Figure 11.8 shows observation for three sites (Kansen, Hakusan and Shimonaga) in Japan.

For various events, the maximum expansion/contraction at an elbow is plotted versus the corresponding expansion/contraction for joints in a straight run. In some cases, the response on the elbow joint was only a tenth of that for a straight pipe joint. However, in other cases, presumably for other angles of incidence, the elbow joint response was three time larger than the straight joint response.

For pipe design purposes, it seems reasonable to use three as the amplification factor for joint openings at bends, relative to the maximum joint opening induced in corresponding straight pipelines. Iwamoto et al. (1985) also measured expansion/contraction for joints adjacent to valve boxes. As shown in Figure 11.9, the behavior is similar to that at elbows in that an amplification factor of three seems appropriate particularly for large ground strains (i.e., when the corresponding straight pipe response is larger).

Similar information from Iwamoto et al. (1985) is presented in Figure 11.10 for joints adjacent to buildings. However, in this case, the amplification factor is as large as 10.

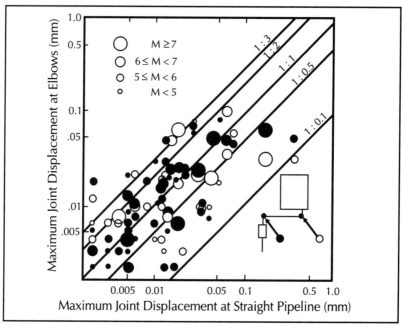

After Iwamoto et al., 1985

■ **Figure 11.8 Observed Joint Displacement and Amplification Factor for Elbows**

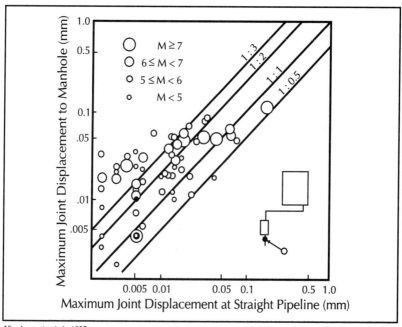

After Iwamoto et al., 1985

■ **Figure 11.9 Observed Joint Displacement and Amplification Factor at Joint Adjacent to Manhole**

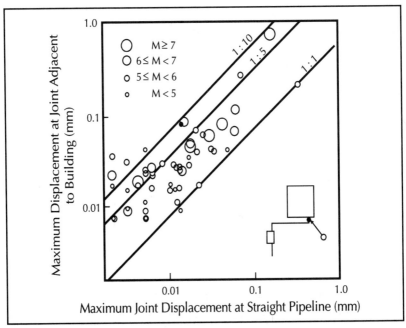

After Iwamoto et al., 1985

■ Figure 11.10 Observed Joint Displacement and Amplification Factor at Joint Adjacent to Buildings

COMPARISON AMONG APPROACHES

This section presents a comparison of three approaches for estimating joint expansion/contraction in a straight run of pipe subject to wave propagation. For the comparison presented in Table 11.1, the peak ground velocity is taken as 0.37 m/sec and the propagation velocity near the ground surface is taken as 240 m/sec which results in a ground strain of 0.00154. The CI pipe has a diameter of about 0.44 m (specifically 0.46 m (18 in) for the Wang model, and 0.41 m (16 in) for the El Hmadi and M. O'Rourke model) and a segment length of 6.1 m. In the Wang approach, the joint axial stiffness is modeled as a linear spring, while in the El Hmadi and M. O'Rourke approach a bi-linear model is used. Specifically for a 16 in diameter with lead caulked joints, the joint stiffness K_1 and K_2 are 3.6×10^5 kN/cm (20.6 kips/in) and 26.5 kN/cm

(1.5 lbs/in), respectively. For comparison purpose, the joint opening from the Wang approach is evaluated separately assuming linear stiffness of 3.6×10^5 kN/cm and 26.5 kN/cm, respectively.

■ Table 11.1 Comparison of Wave Propagation Response of Straight Segmented Pipelines

Item	Eqn 11.1 (Max.)	Wang (Max.)		El Hmadi & O'Rourke (Average)
		$K_j = 3.6310^5$ kN/cm	$K_j = 26.5$ kN/cm	
Ground Strain	0.00154	0.00154		0.00154
Pipe Strain	-	1.2×10^{-3}	4.0×10^{-5}	-
Joint Opening (cm)	0.92	0.21	0.8	0.83 Average 2.0 1% Exceedence 3.4 0.1% Exceedence

The average joint opening from the El Hmadi and M. O'Rourke approach matches reasonably well with the results from Equation 11.1 and with the Wang approach for $K_j = 26.5$ kN/cm. However, when the initial joint stiffness of 3.6×10^5 kN/cm is used in Wang's approach, the joint opening is much smaller and the axial strain in the pipe segments is much larger as shown in Table 11.1. This illustrates that care must be taken when attempting to model bi-linear behavior (in this case the axial stiffness of CI pipe joints) with a linear model.

Based upon a joint depth (d_p in Figure 4.11) of 11.5 cm for a 16 to 18 in diameter pipe, one would not expect damage due to joint pull-out for a joint opening in the range of 0.8 to 0.92 cm since, as mentioned in Section 4.2.1, the normalized joint displacement is less than about a half (0.92/11.5<<0.5). However, using the El Hmadi and M. O'Rourke approach, the displacement for one in a thousand joints is 3.4 cm (normalized displacement = 3.4/11.5=0.3) and hence, some leakage (about one in 10,000 (1/10×1/1000) joints from Figure 4.11) is expected.

An advantage of the Wang approach is that an estimate of the pipe strain is provided. However, based upon the above discussion, the expected tensile strain in the segments (4.0×10^{-5} strain corresponding to 0.8 cm joint opening) is less than the yield strain for a CI pipe of about 2.0×10^{-3} from Chapter 4. Hence, although pipe strain is provided, it is unlikely that the pipe segment failure mode governs.

E F F E C T S   O F   L I Q U E F I E D   S O I L

The response of both segmented and continuous pipe to PGD was discussed in previous chapters. Also the uplifting response of pipe surrounded by liquefied soil was treated in Chapter 7. In this section, the somewhat special case of a buried pipeline subject to sloshing action due to liquefaction of a subsurface layer is addressed. Note that 69 breaks to segmented water mains within the Marina District during the 1989 Loma Prieta event have been attributed to this mechanism.

Laboratory experiments by Nishio et al. (1987, 1989) provided an observational basis for the phenomenon. They analyzed the dynamic response of a continuous pipeline surrounded by non-liquefied soil but underlain by a liquefied layer. Figure 11.11 shows their test set up. The pipe is 13 mm (0.51 in) in diameter and made of poly-carbonate.

The input motion at the base of the model was a sinusoidal wave with a frequency of 5 Hz and a peak acceleration of about 150 gal. Accelerations throughout the model were comparable to the base acceleration up to the time of liquefaction of the loose sand deposit. After liquefaction, the acceleration of the non-liquefied surface layer directly over the loose sand deposit was amplified by a factor of roughly 2, relative to the input acceleration. Al-

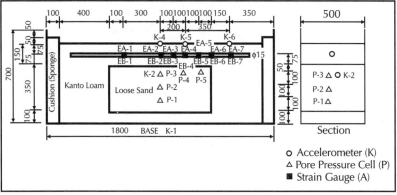

After Nishio et al., 1987

■ Figure 11.11 Configuration of Model and Locations of Gauges

though the base motion was horizontal (parallel to the pipe axis), the resulting pipe motion after liquefaction had a significant vertical component and the vertical motion was asymmetric about the center of the liquefied deposit (i.e., a sloshing type of response). Figures 11.12 and 11.13 show the distributions of pipe bending strains and axial strains at two times during the test (both after the onset of liquefaction of the subsurface deposit).

Note that the amplitude of the bending strain is roughly an order of magnitude larger than the axial strain. The axial strain is largest near the edge of the liquefaction deposit, while the bending strain is largest about a quarter of the distance from the edge. Note that a point of counter flexure exists near the center of the model, confirming the asymmetric nature of the response (e.g. left portion moving up while right portion moves down).

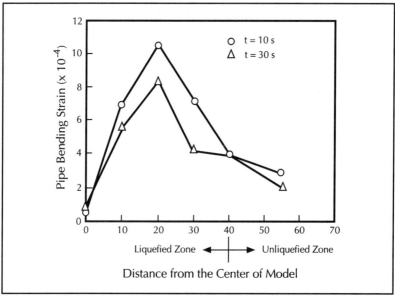

After Nishio et al., 1987

■ Figure 11.12 Distribution of Peak Pipe Bending Strain

Ishihara (1985) presents a relation between acceleration and thicknesses of non-liquefied surface layer and liquefied sublayer in Figure 11.14 based on the damage observation during past earthquakes. This observation suggests that for a given value of peak acceleration, the thinner non-liquefied surface layer, the more serious the damage.

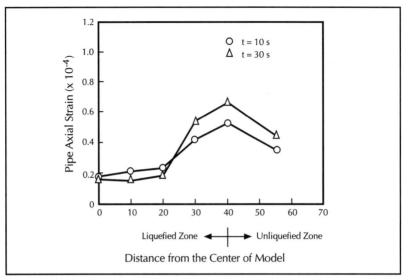

■ Figure 11.13 **Distribution of Peak Pipe Axial Strain**

Considering date from a wide range of earthquake and site condition, Youd and Garris (1995) evaluate and verify Ishihara's criteria. Their study suggests that the thickness bounds proposed by Ishihara appear valid for sites not susceptible to lateral spread, but not valid for site susceptible to lateral spread.

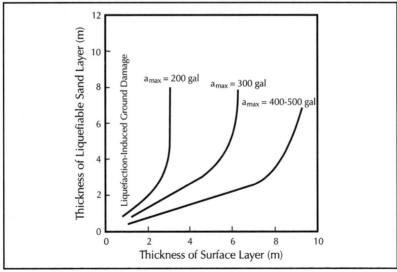

■ Figure 11.14 **Ishihara's Proposed Boundary Curves for Site Identification of Liquefaction-Induced Damage**

COUNTERMEASURES TO MITIGATE SEISMIC DAMAGE

As indicated previously, seismic damage to buried pipeline is due to either PGD and/or wave propagation. Prior studies have suggested various methods to mitigate against seismic damage to pipelines. These include high strength or high ductility materials for the pipelines themselves, the use of joints with enhanced expansion/contraction or rotation capability, various methods to isolate the pipeline from ground movements, various methods to reduce the amount of ground movement, and finally, avoiding or rerouting pipes around areas particularly susceptible to damaging ground movements.

This chapter will briefly describe each of these techniques and identify situations in which they may be particularly effective.

12.1 ROUTING AND RELOCATION

This technique involves simply avoiding areas which are susceptible to large ground movements. It is comparatively easy to implement during the initial design (i.e., route selection) stage for a new pipeline, but can also be used for an existing line. For example, Bukovansky et al. (1985) report on the relocation of a 66 cm (26 in.) natural gas pipeline which was subject to a landslide hazard. About 365 m (1200 ft) of line was relocated at a direct cost of roughly 1 million dollars. As one might expect, relocation of an existing line typically requires temporary suspension of service.

This method would typically be more effective for the PGD hazard such as landslides or areas susceptible to liquefaction. It could also be used for the fault crossing hazard if the end points for the line are both on the same side of the active fault. That is, relocation tends to be more effective when hazard exists only in an isolated area which can be avoided. Routing and relocation tends to be less effective for wave propagation damage, since this hazard typically exists over much larger areas. Finally, routing and relocation would typically be easier to implement for transmission pipe, for which there may be a number of options in terms of route selection, but more difficult for distribution pipe. For example, if natural gas service is needed along a given street, alternate locations may be severely limited.

ISOLATION FROM DAMAGING GROUND MOVEMENT

As described above, routing and relocation involve alternate locations (i.e., realignment in the horizontal plane). When such alternate locations are unavailable, impractical or cost prohibitive, isolation techniques can be used to mitigate against seismic damage to pipelines. In this case, the pipeline traverses the hazardous area but is isolated from the effects of large ground movements by realignment in the vertical direction. A classic example is the placement of the Trans-Alaskan pipeline on above ground "goal post" type supports at fault crossing locations. That is, for strike-slip faults there is enough "rattle-space" between the uprights that the potential fault movement can be accommodated without overstressing the pipe. This method can be used for most types of PGD hazards; however, proper implementation often requires a low-friction sliding surface between the pipe and its horizontal supporting member.

For certain PGD hazards, the same objectives can be obtained by directional drilling technology. In this case, the pipe is isolated from potential damage by being located below the hazardous area. Directional drilling can be used for the landslide hazard as well as the liquefaction hazard. It is particularly attractive at river crossings which may be susceptible to liquefaction induced PGD of the bank. However, this technique cannot be used effectively at faults, since it is not possible to place the pipe "below" the fault. A third mitigation approach, within this isolation class, involves orientation of the pipe so as to reduce the potential for damage. As shown in Chapters 6 and 7, the potential for damage to a continuous pipe subject to PGD is reduced as the line is orientated perpendicular to the direction of ground movement (i.e., transverse PGD as opposed to longitudinal PGD). Similarly, as shown in Chapter 8, a continuous pipe subject to the fault crossing hazard should be orientated such that the fault movement places the line in tension as opposed to compression. Theoretically, the optimum situation corresponds to the pipe at right angles to the fault. However, due to the difficulty associated with establishing the actual orientation of the fault line in a horizontal plane, an angle, β, of about 60° is recommended.

For segmented pipe, the preferred orientations are the same. That is, as shown in Chapter 9, transverse PGD is preferable to longitudinal PGD particularly if the joints are flexible. Also, at fault crossings, an angle close to 90° is marginally better for typical joint types.

REDUCTION OF GROUND MOVEMENTS

These mitigation techniques involve various types of field treatments to reduce the potential for lateral spreading. The methods include increasing the density and strength of sand, lowering the ground water level and increasing the dissipation of pore water pressure. For example, Miyajima et al. (1992) proposed a vertical gravel drain system along the pipeline right-of-way which reduces the maximum pore water pressures. Fujii et al. (1992) suggest sand and compaction as a technique to increase soil density and strength, and thereby reduce the potential for liquefaction. Iwatate et al. (1988) performed experiments on buried culverts which drain ground water away from the pipeline. Finally, one could replace liquefaction soils in the vicinity of the pipe with nonliquifiable materials such as gravel to reduce the potential for liquefaction.

These field treatment methods tend to be practical only when the spatial extent of the liquefied soil deposits is limited and the liquifable soil layer is relatively close to the ground surface. They are less practical and cost effective for the typical landslide hazard. In a retrofit situation, they would not necessarily result in service disruption if done carefully.

12.4

HIGH STRENGTH MATERIALS

For continuous pipe, improved seismic performance results, as one might expect, from the use of stronger materials (i.e., higher nominal yield stress) and larger pipe wall thickness. For example, Table 6.3 shows that the higher strength X-70 pipe can accommodate more longitudinal PGD than a corresponding Grade B pipe. Similar improved performance for higher strength pipe subject to transverse PGD and fault crossing are shown in Figure 7.18 (reduced tensile strain) and Figure 8.11 (larger allowable fault offset), respectively. Similarly, Table 6.3, Figure 7.15 and Figure 8.4 show improved performance for thicker wall pipe subject to longitudinal PGD, transverse PGD and fault crossing, respectively. Note that this improved performance is for steel pipe with electric arc welded butt joints. The performance of steel pipe with slip joints, rivet joints or oxy-acetylene welds is expected to be poorer.

Another mitigation option involves reducing the load as opposed to increasing the strength. As shown in Section 6.2, the axial strain induced in a continuous pipe by longitudinal PGD is an increasing function of the pipe burial parameter β_p as defined in Equation 6.5.

$$\beta_p = \frac{\mu\gamma H}{t} \qquad (12.1)$$

Hence, axial strain can be reduced by using the smallest possible burial depth (H), using low density backfill (γ), and/or using coatings which reduce the coefficient of friction at the soil-pipe interface (μ).

These mitigation options can be easily used for design of new pipelines and also for retrofit of key sections of existing pipe. Of course, in the retrofit situation, replacement would in all likelihood result in disruption of service.

FLEXIBLE MATERIALS AND JOINTS

It has long been argued that the use of more flexible materials tends to improve the seismic performance of buried pipeline. The expected benefits of flexible materials are shown in Figure 1.2. Note that there are fewer repairs per unit length for ductible iron (DI) and polyethylene (PE) pipe, at a given MMI level, than for more brittle materials such as asbestos cement (AC), cast iron (CI), concrete (Conc), and polyvinyl chloride (PVC). As shown in Figures 1.5 and 1.6, one similarly expects improved seismic performance for flexible pipe materials, such as arc-welded steel and ductible iron pipe, when subject to various PGD hazards.

For segmented pipe, Isenberg and Richardson (1989), Ballantyne (1992) and Wang (1994) have suggested the use of flexible joints for pipeline subject to the PGD hazard. Figure 12.1 presents sketches of various joint types while Table 12.1 lists the expected deformation capacity, based upon information provided by Singhal (1984), Isenberg and Richardson, and Akiyoshi et al. (1994). However, as explained in Section 6.3, expansion joints

■ **Table 12.1 Deformation Capacity of Flexible Joints**

Item	Pull-Out	Rotation	Note
Mechanical Joint	3 cm	5°	
Locked Mechanical Joint	<1 cm	5°	Slight Expansion
Restrained Mechanical Joint	5 cm	5°	S-Type Joint (Japan)
Tyton Joint	3 cm	3° - 5°	Vary with Pipe Diameter
Flange-locked Joint	3 cm	5°	
TR FLEX Telescoping Sleeve	2 D		D—Pipe Diameter
Restrained Expansion Joint	25 cm	5°	
XTRA FLEX Coupling		20°	
Ball Joint		15°	

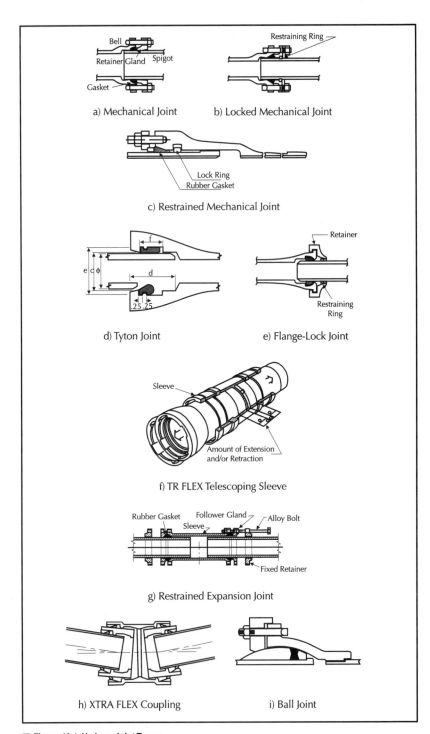

a) Mechanical Joint

b) Locked Mechanical Joint

c) Restrained Mechanical Joint

d) Tyton Joint

e) Flange-Lock Joint

f) TR FLEX Telescoping Sleeve

g) Restrained Expansion Joint

h) XTRA FLEX Coupling

i) Ball Joint

■ Figure 12.1 Various Joint Types

need to be used with caution. For example, if an expansion joint is placed at only one end of a lateral spread zone, the strain in a continuous pipe induced by longitudinal PGD would actually be larger than that for a pipe with no expansion joints.

For localized abrupt PGD offsets, such as at a fault crossing, Ford (1983) suggests, as shown in Figure 12.2, the use of rotationally flexible ball type joints in combination with an expansion joint. This combined rotation and extension flexibility could also be useful at other locations when differential movements are expected. Examples of such locations include the margins of areas with variable subsurface conditions and inlet/outlet to stiff structures such as tanks and buildings.

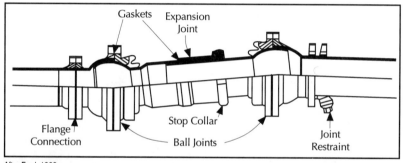

After Ford, 1983

■ Figure 12.2 Combined Ball and Expansion Joint

R E F E R E N C E S

Akiyoshi, T., Fuchida, K., Shirinashihama, S., and Tsutsumi, T., (1994), "Effectiveness of Anti-Liquefaction Techniques for Buried Pipelines," *Journal of Pressure Vessel Technology*, Vol. 116, August, pp. 261-266.

American Society of Civil Engineers (ASCE), (1984), *Guidelines for the Seismic Design of Oil and Gas Pipeline Systems*, Committee on Gas and Liquid Fuel Lifeline, ASCE.

Ando, H. Sato, S., and Takagi, N., (1992), "Seismic Observation of a Pipeline Buried at the Heterogeneous Ground," *Proceedings of the Tenth World Conference on Earthquake Engineering*, Balkema, Rotterdam, pp. 5563-5567.

Applied Technology Council, (1985), *Earthquake Damage Evaluation Data for California*, ATC-13, Redwood City, California.

Ariman, T. and Lee, B.J., (1989), "On Beam Mode of Buckling of Buried Pipelines," *Proceedings of the Second U.S.-Japan Workshop on Liquefaction, Large Ground Deformation and Their Effects on Lifelines*, Buffalo, New York, Technical Report NCEER-89-0032, Multidisciplinary Center for Earthquake Engineering Research, Buffalo, New York, pp. 401-412.

Ariman, T. and Lee, B.J., (1991), "Tension/Bending Behavior of Buried Pipelines Under Large Ground Deformation in Active Faults," *Proceedings of the Third U.S. Conference on Lifeline Earthquake Engineering*, Technical Council on Lifeline Earthquake Engineering, Monograph No. 4, ASCE, pp. 226-233.

Audibert, J.M.E. and Nyman, K.J., (1977), "Soil Restraint Against Horizontal Motion of Pipes," *Journal of Geotechnical Engineering Division*, ASCE, Vol. 103, No. GT10, pp. 1119-1142.

Ayala, G. and O'Rourke, M., (1989), *Effects of the 1985 Michoacan Earthquake on Water Systems and Other Buried Lifelines in Mexico*, Technical Report NCEER-89-0009, Multidisciplinary Center for Earthquake Engineering Research, Buffalo, New York.

Ballantyne, D., (1992), "Thoughts on a Pipeline Design Standard Incorporating Countermeasure for Permanent Ground Deformation," *Proceedings of the Fourth Japan-U.S. Workshop on Earthquake Resistant Design of Lifeline Facilities and Countermeasures for Soil Liquefaction*, Honolulu, Hawaii, Technical Report NCEER-92-0019, Multidisciplinary Center for Earthquake Engineering Research, Buffalo, New York, pp. 875-887.

Barenberg, M.E., (1988), "Correlation of Pipeline Damage with Ground Motions," *Journal of Geotechnical Engineering*, ASCE, June, Vol. 114, No. 6, pp. 706-711.

Bartlett, S.F. and Youd, T.L., (1992), *Empirical Analysis of Horizontal Ground Displacement Generated by Liquefaction-induced Lateral Spreads*, Technical Report NCEER-92-0021, Multidisciplinary Center for Earthquake Engineering Research, Buffalo, New York.

Bouabid, J. and O'Rourke, M.J., (1994), "Seismic Vulnerability of Concrete Pipelines," *Proceedings of the Fifth U.S. National Conference on Earthquake Engineering*, Chicago, Illinois, EERI, July, Vol. IV, pp. 789-798.

Bouabid, J., (1995), "Behavior of Rubber Gasketed Concrete Pipe Joints During Earthquakes," Ph.D. Thesis, Rensselaer Polytechnic Institute, December.

Brockenbrough, R.L., (1990), "Strength of Bell-and-Spigot Joints," *Journal of Structural Engineering*, Vol. 116, No. 7, pp. 1983-1991.

Bukovansky, M., Greenwood, J.H., Major, G., (1985), "Maintaining Natural Gas Pipeline in Active Landslides," *Advances in Underground Pipe Engineering*, ASCE, (J.K. Jeyapalan, Editor), pp. 438-448.

Cai, J., Liu, X., Hou, Z., and Chen, G., (1992), "Experimental Research on the Liquefaction Response of Oil Supply Pipeline," *Proceedings of the Second International Symposium on Structural Technique of Pipeline Engineering*, Beijing, pp. 284-291.

Campbell, K.W. and Bozorgnia, Y., (1994), "Near-Source Attenuation of Peak Horizontal Acceleration from Worldwide Accelerograms Recorded from 1957 to 1993," *Proceedings of the Fifth U.S. National Conference on Earthquake Engineering*, Chicago, Illinois, July 10-14, EERI, Vol. III, pp. 283-292.

Dieckgrafe, R.E., (1976), *Personal Communication*, Transmitted from the Manager, Texaco Puerto Cortes Refinery, Honduras, November.

Dobry, R. and Baziar, M.H., (1990), "Evaluation of Ground Deformation Caused by Lateral Spreading," *Proceedings of the Third Japan-U.S. Workshop on Earthquake Resistant Design of Lifeline Facilities and Countermeasures for Soil Liquefaction*, San Francisco, California, December 17-19, Technical Report NCEER-91-0001, Multidisciplinary Center for Earthquake Engineering Research, Buffalo, New York, pp. 209-224.

Eguchi, R.T., (1983), "Seismic Vulnerability Models for Underground Pipes," *Earthquake Behavior and Safety of Oil and Gas Storage Facilities, Buried Pipelines and Equipment*, PVP-77, ASME, New York, June, pp. 368-373.

Eguchi, R., (1991), *Early Post-Earthquake Damage Detection for Underground Lifelines*, Final Report to the National Science Foundation, Dames and Moore P.C., Los Angeles, California.

Eidinger, J.M., Maison, B., Lee, D., and Lau, B., (1995), "East Bay Municipal Utility District Water Distribution Damage in 1989 Loma Prieta Earthquake," *Proceedings of the Fourth U.S. Conference on Lifeline Earthquake Engineering*, ASCE, Technical Council on Lifeline Earthquake Engineering, Monograph No. 6, ASCE, August, pp. 240-247.

El Hmadi, K. and O'Rourke, M.J., (1989), *Seismic Wave Effects on Straight Jointed Buried Pipeline*, Technical Report NCEER-89-0022, Multidisciplinary Center for Earthquake Engineering Research, Buffalo, New York.

El Hmadi, K. and O'Rourke, M.J., (1990), "Seismic Damage to Segmented Buried Pipelines," *Earthquake Engineering and Structural Dynamics*, May, Vol. 19, No. 4, pp. 529-539.

Federal Emergency Management Agency (FEMA), (1991), *Seismic Vulnerability and Impact of Disruption of Lifelines in the Conterminous United States*, Applied Technology Council, ATC-25, FEMA 224, September.

Flores-Berrones, R. and O'Rourke, M., (1992), "Seismic Effects on Underground Pipelines Due to Permanent Longitudinal Ground Deformations," *Proceedings of the Fourth Japan-U.S. Workshop on Earthquake Resistant Design of Lifeline Facilities and Countermeasures for Soil Liquefaction*, Honolulu, Hawaii, Technical Report NCEER-92-0019, Multidisciplinary Center for Earthquake Engineering Research, Buffalo, New York, pp. 465-480.

Ford, D.B., (1983), "Joint Design for Pipelines Subjected to Large Ground Deformations," *Earthquake Behavior and Safety of Oil and Gas Storage Facilities, Buried Pipelines and Equipment*, PVP-77, ASME, New York, June, pp. 160-165.

Fujii, Y., Ohtomo, K., Arai, H., and Hasegawa, H., (1992), "The State of the Art in Mitigation of Liquefaction for Lifeline Facilities in Japan," *Proceedings of the Fourth Japan-U.S. Workshop on Earthquake Restraint Design of Lifeline Facilities and Countermeasures for Soil Liquefaction*, Honolulu, Hawaii, Technical Report NCEER-92-0019, Multidisciplinary Center for Earthquake Engineering Research, Buffalo, New York, pp. 889-909.

G&E, (1994), *NIBS Earthquake Loss Estimation Methods, Technical Manual, (Water Systems)*, May.

Goodling, E.C., (1983), "Buried Piping — An Analysis Procedure Update," *Earthquake Behavior and Safety of Oil and Gas Storage Facilities, Buried Pipelines and Equipment*, PVP-77, ASME, New York, June, pp. 225-237.

Gregor, N.J., (1995), *The Attenuation of Strong Ground Motion Displacements*, Earthquake Engineering Research Center, Report Number UCB/EERC-95/02, University of California at Berkeley, June.

Hall, J.F., (1995), "Northridge Earthquake of January 17, 1994, Reconnaissance Report," *Earthquake Spectra*, EERI, April.

Hall, W. and Newmark, N., (1977), "Seismic Design Criteria for Pipelines and Facilities," *Current State of Knowledge of Lifeline Earthquake Engineering*, ASCE, New York, pp. 18-34.

Hamada, M., Yasuda, S., Isoyama, R., and Emoto, K., (1986), *Study on Liquefaction Induced Permanent Ground Displacements*, Association for the Development of Earthquake Prediction, Japan, 87p.

Hamada, M., Towhata, T., Yasuda, S., and Isoyama, R., (1987), "Study of Permanent Ground Displacement Induced by Seismic Liquefaction," *Computers and Geotechnics*, Elsevier Applied Science Publishers, Vol. 4, No. 4, pp. 197-220.

Hamada, M., (1989), "Damage to Buried Lifelines Due to Liquefaction-Induced Ground Displacements," *Proceedings of the Third U.S.-Japan Workshop on Earthquake Disaster Prevention for Lifeline Systems*.

Hamada, M. and O'Rourke, T.D., (1992), "Large Ground Deformations and Their Effects on Lifelines," *Japanese Case Studies of Liquefaction and Lifeline Performance During Past Earthquakes*, Technical Report NCEER-92-0001, Multidisciplinary Center for Earthquake Engineering Research, Buffalo, New York.

Hansen, W.R., (1971), *Effects at Anchorage, The Great Alaska Earthquake of 1964*, National Academy of Sciences, Washington, DC, pp. 289-358.

Haskell, N.A., (1953), "The Dispersion of Surface Waves in Multilayered Media," *Bulletin of the Seismological Society of America*, Vol. 43, No. 1, pp. 17-34.

Heubach, W.F., (1995), "Seismic Damage Estimation for Buried Pipeline Systems," *Proceedings of the Fourth U.S. Conference on Lifeline Earthquake Engineering*, Technical Council on Lifeline Earthquake Engineering, Monograph No. 6, ASCE, pp. 312-319.

Hobbs, R.E., (1981), "Pipeline Buckling Caused by Axial Loads," *Journal of Constructional Steel Research*, January, Vol. 1, No. 2, pp. 2-10.

Honegger, D.G. and Eguchi, R.T., (1992), *Determination of Relative Vulnerabilities to Seismic Damage for San Diego County Water Authority (SDCWA) Water Transmission Pipelines*, October.

Honegger, D.G., (1995), "An Approach to Extend Seismic Vulnerability Relationships for Large Diameter Pipelines," *Proceedings of the Fourth U.S. Conference on Lifeline Earthquake Engineering*, Technical Council on Lifeline Earthquake Engineering, Monograph No. 6, ASCE, pp. 320-327.

Hou, Z., Cai, J., and Liu, X., (1990), "Response Calculation of Oil Pipeline Subjected to Permanent Ground Movement Induced by Soil Liquefaction," *Proceedings of the China-Japan Symposium on Lifeline Earthquake Engineering*, Beijing, China, pp. 107-114.

Howard, J.H., (1968), "Recent Deformation at Buena Vista Hills, California," *American Journal of Science*, November, Vol. 266, pp. 737-757.

Isenberg, J. and Escalante, L.E., (1984), "Damage to Lifelines," *Reconnaissance Report, Coalinga, California, Earthquake of May 2, 1982*, EERI, January, pp. 227-247.

Isenberg, J. and Richardson, E., (1989), "Countermeasures to Mitigate Damage to Pipelines," *Proceedings of the Second U.S.-Japan Workshop on Liquefaction, Large Ground Deformation and Their Effects on Lifelines*, Buffalo, New York, Technical Report NCEER-89-0032, Multidisciplinary Center for Earthquake Engineering Research, Buffalo, New York, pp. 468-482.

Ishihara, K., (1985), "Stability of Natural Deposits During Earthquake," *Proceedings of the Eleventh International Conference on Soil Mechanics and Foundation Engineering*, Vol. 1, pp. 321-376.

Isoyama, R. and Katayama, T., (1982), *Reliability Evaluation of Water Supply Systems During Earthquakes*, Report of the Institute of Industrial Science, University at Tokyo, February, Vol. 30, No. 1.

Iwamoto, T., Wakai, N., and Yamaji, T., (1984), "Observation of Dynamic Behavior of Buried Ductile-Iron Pipelines During Earthquakes," *Eighth World Conference on Earthquake Engineering*, San Francisco, California, Vol. VII, pp. 231-238.

Iwamoto, T., Yamamura, Y., and Hojo, S., (1985), "Observations and Analyses at the Bend and Tee Portions of Buried Ductile Pipelines During Earthquakes," *Seismic Performance of Pipelines and Storage Tanks*, PVP-98-4, ASME, pp. 69-79.

Iwatate, T., Ohtomo, K., Tohma, J., and Nozawa, Y., (1988), "Liquefaction Disaster Mitigation for Underground Structures," *Proceedings of the First Japan-U.S. Workshop on Liquefaction, Large Ground Deformation and Their Effects on Lifeline Facilities*, Tokyo, Japan, November, pp. 143-151.

Japan Gas Association, (1982), *Specifications for Seismic Design of High Pressure Gas Pipelines*.

Japan Road Association, (1990), *Japan Road Association Specifications for Highway Bridges, Part V, Earthquake Resistant Design*, Tokyo, Japan.

Jibson, R.W. and Keefer, D.K., (1993), "Analysis of the Seismic Origin of Landslides: Examples from the New Madrid Seismic Zone," *Geological Society of America Bulletin*, April, Vol. 105, pp. 521-536.

Joyner, W.B. and Boore, D.M., (1981), "Peak Horizontal Acceleration and Velocity from Strong-Motion Records Including Records from the 1979 Imperial Valley, California, Earthquake," *Bulletin of the Seismological Society of America*, December, Vol. 71, No. 6, pp. 2011-2038.

Kachadoorian, R., (1976), "Earthquake: Correlation Between Pipeline Damage and Geologic Environment," *Journal of the American Water Works Association*, March, pp. 165-167.

Katayama, T., Kubo, K., and Sato, N., (1975), "Earthquake Damage to Water and Gas Distribution Systems," *Proceedings of the U.S. National Conference on Earthquake Engineering*, Oakland, California, EERI, pp. 396-405.

Kamiyama, M., O'Rourke, M.J., and Flores-Berrones, R., (1992), *A Semi-Empirical Analysis of Strong-Motion Peaks in Terms of Seismic Source, Propagation Path and Local Site Conditions*, Technical Report NCEER-92-0023, Multidisciplinary Center for Earthquake Engineering Research, Buffalo, New York.

Kennedy, R.P., Chow, A.W., and Williamson, R.A., (1977), "Fault Movement Effects on Buried Oil Pipeline," *Journal of the Transportation Engineering Division*, ASCE, May, Vol. 103, No. TE5, pp. 617-633.

Kobayashi, T., Nakane, H., Suzuki, N., and Ishikawa, M., (1989), "Parametric Study on Flexibility of Buried Pipeline Subject to Large Ground Displacement," *Proceedings of the Second U.S.-Japan Workshop on Liquefaction, Large Ground Deformation and Their Effects on Lifelines*, Buffalo, New York, Technical Report NCEER-89-0032, Multidisciplinary Center for Earthquake Engineering Research, Buffalo, New York, pp. 348-362.

Krathy, R.G. and Salvadori, M.G., (1978), *Strength and Dynamic Characteristics of Gasket-Jointed Concrete Water Pipelines*, Weidlinger Associates, Grant Report, No. 5.

Kubo, K., (1974), "Behavior of Underground Water Pipes During an Earthquake," *Proceedings of the Fifth World Conference on Earthquake Engineering*, Rome, Italy, pp. 569-578.

Kyriakides, S., Yun, H.D., and Yew, C.H., (1983), "Buckling of Buried Pipelines Due to Large Ground Movements," *Earthquake Behavior and Safety of Oil and Gas Storage Facilities, Buried Pipelines and Equipment*, PVP-77, ASME, New York, June, pp. 140-150.

Liu, X. and Hou, Z., (1991), "Seismic Risk of Water Supply Network," *Proceedings of the Third U.S. Conference on Lifeline Earthquake Engineering*, Technical Council on Lifeline Earthquake Engineering, Monograph No. 4, ASCE, pp. 552-561.

Liu, X. and O'Rourke, M., (1997a), "Seismic Ground Strain at Sites with Variable Subsurface Conditions," *Computer Methods and Advances in Geomechanics*, (J.X. Yuan, Editor), A.A. Balkema, pp. 2239-2244.

Liu, X. and O'Rourke, M., (1997b), "Behavior of Continuous Pipeline Subject to Transverse PGD," *Journal of Earthquake Engineering and Structural Dynamics*, Vol. 26, pp. 989-1003.

Mabey, M.A., (1992), *Prediction of Displacement Due to Liquefaction Induced Lateral Spreading*, Doctoral Dissertation, Department of Civil Engineering, Brigham Young University, Provo, Utah, Technical Report CEG-92-02.

Manson, M., (1908), *Reports on an Auxiliary Water Supply System for Fire Protection for San Francisco, California*, Report of Board of Public Works, San Francisco, California.

Marek, P.J. and Daniels, J.H., (1971), "Behavior of Continuous Crane Rails," *Journal of the Structural Division*, ASCE, April, Vol. 97, No. ST4, pp. 1081-1095.

Markov, I., Grigoriu, M., and O'Rourke, T., (1994), *An Evaluation of Seismic Serviceability of Water Supply Networks with Application to San Francisco Auxiliary Water Supply System*, Technical Report NCEER-94-0001, Multidisciplinary Center for Earthquake Engineering Research, Buffalo, New York.

Matsumoto, H., Sasaki, Y., and Kondo, M., (1987), "Coefficient of Subgrade Reaction on Pile in Liquefied Ground," *Proceedings of the Second National Conference on Soil Mechanics and Foundation Engineering*, pp. 827-828 (in Japanese).

McCaffrey, M.A. and O'Rourke, T.D., (1983), "Buried Pipeline Response to Reverse Faulting During the 1971 San Fernando Earthquake," *Earthquake Behavior and Safety of Oil and Gas Storage Facilities, Buried Pipelines and Equipment*, PVP-77, ASME, New York, June, pp. 151-159.

McNorgan, J.D., (1989), "Relieving Seismic Stresses Locked in Gas Pipelines," *Proceedings of the Second U.S.-Japan Workshop on Liquefaction, Large Ground Deformation and Their Effects on Lifelines*, Buffalo, New York, Technical Report NCEER-89-0032, Multidisciplinary Center for Earthquake Engineering Research, Buffalo, New York, pp. 363-369.

Meyersohn, W.D., (1991), *Analytical and Design Considerations for the Seismic Response of Buried Pipelines*, Thesis, Graduate School of Cornell University, January.

Miyajima, M. and Kitaura, M., (1989), "Effects of Liquefaction-Induced Ground Movement on Pipeline," *Proceedings of the Second U.S.-Japan Workshop on Liquefaction, Large Ground Deformation and Their Effects on Lifelines*, Buffalo, New York, Technical Report NCEER-89-0032, Multidisciplinary Center for Earthquake Engineering Research, Buffalo, New York, pp. 386-400.

Miyajima, M. and Kitaura, M., (1991), "Experiments on Soil Spring Constants During Liquefaction," *Proceedings of the Second International Conference on Recent Advances in Geotechnical Earthquake Engineering and Soil Dynamics*, March 11-15, St. Louis, Missouri.

Miyajima, M., Yoshida, M., and Kitaura, M., (1992), "Small Scale Tests on Countermeasures Against Liquefaction for Pipelines Using Gravel Drain System," *Proceedings of the Fourth Japan-U.S. Workshop on Earthquake Resistant Design of Lifeline Facilities and Countermeasures for Soil Liquefaction*, Honolulu, Hawaii, NCEER-92-0019, Multidisciplinary Center for Earthquake Engineering Research, Buffalo, New York, pp. 381-391.

Moncarz, P.D., Shyne, J.C., and Derbalian, G.K., (1987), "Failures of 108-inch Steel Pipe Water Main," *Journal of Performance of Construction Facilities*, ASCE, Vol. 1, No.3, pp. 168-187.

Newmark, N.M., (1965), "Effects of Earthquakes on Dams and Embankments," *Geotechnique*, Vol. 15, No. 2, pp. 139-160.

Newmark, N.M., (1967), "Problems in Wave Propagation in Soil and Rocks," *Proceedings of the International Symposium on Wave Propagation and Dynamic Properties of Earth Materials*, University of New Mexico Press, pp. 7-26.

Newmark, N.M. and Hall, W.J., (1975), "Pipeline Design to Resist Large Fault Displacement," *Proceedings of the 1975 U.S. National Conference on Earthquake Engineering*, Ann Arbor, Michigan, pp. 416-425.

National Institute of Building Sciences (NIBS), (1996), *Development of a Standardized Earthquake Loss Estimation Methodology*, Vol. II, Risk Management Solutions, Inc., February 15.

Nishio, N., Ukaji, T., Tsukamoto, K., and Ishita, O., (1983), "Model Experiments on the Behavior of Buried Pipelines During Earthquakes," *Earthquake Behavior and Safety of Oil and Gas Storage Facilities, Buried Pipelines and Equipment*, PVP-77, ASME, New York, June, pp. 263-272.

Nishio, N., Hamura, A., and Sase, T., (1987), *Model Experiments on the Seismic Behavior of Buried Pipeline During Liquefaction of Ground (II), —In Consideration of Similarity Condition—*, Technical Report of R & D Institute of Tokyo Gas Co. Ltd., Vol. 32, (In Japanese).

Nishio, N., (1989), "Dynamic Strains in Buried Pipelines Due to Soil Liquefaction," *Earthquake Behavior of Buried Pipelines, Storage, Telecommunication, and Transportation Facilities*, PVP-162, ASME, pp. 83-88.

Orense, R. and Towhata, I., (1992), "Prediction of Liquefaction-Induced Permanent Ground Displacements: A Three-Dimensional Approach," *Proceedings of the Fourth Japan-U.S. Workshop on Earthquake Resistant Design of Lifeline Facilities and Countermeasures for Soil Liquefaction*, Honolulu, Hawaii, Technical Report NCEER-92-0019, Multidisciplinary Center for Earthquake Engineering Research, Buffalo, New York, pp. 335-349.

O'Rourke, M. and Wang, L.R.L., (1978), "Earthquake Response of Buried Pipelines," *Proceedings of the Special Conference on Earthquake Engineering and Soil Dynamics*, Pasadena, California, ASCE, pp. 720-731.

O'Rourke, M.J., Bloom, M.C., and Dobry. R., (1982), "Apparent Propagation Velocity of Body Waves," *Earthquake Engineering and Structural Dynamics*, Vol. 10, pp. 283-294.

O'Rourke, M.J., Castro, G., and Hossain, I., (1984), "Horizontal Soil Strain Due to Seismic Waves," *Journal of Geotechnical Engineering*, ASCE, September, Vol. 110, No. 9, pp. 1173-1187.

O'Rourke, M.J. and El Hmadi, K.E., (1988), "Analysis of Continuous Buried Pipelines for Seismic Wave Effects," *Earthquake Engineering and Structural Dynamics*, Vol. 16, pp. 917-929.

O'Rourke, M.J., (1989), "Approximate Analysis Procedures for Permanent Ground Deformation Effects on Buried Pipelines," *Proceedings of the Second U.S.-Japan Workshop on Liquefaction, Large Ground Deformation and Their Effects on Lifelines*, Buffalo, New York, Technical Report NCEER-89-0032, Multidisciplinary Center for Earthquake Engineering Research, Buffalo, New York, pp. 336-347.

O'Rourke, M.J. and Ayala, G., (1990), "Seismic Damage to Pipeline: Case Study," *Journal of Transportation Engineering*, ASCE, March/April, Vol. 116, No. 2, pp. 123-134.

O'Rourke, M. and Nordberg, G., (1991), "Analysis Procedures for Buried Pipelines Subject to Longitudinal and Transverse Permanent Ground Deformation," *Proceedings of the Third Japan-U.S. Workshop on Earthquake Resistant Design of Lifeline Facilities and Countermeasures for Soil Liquefaction*, San Francisco, California, Technical Report NCEER-91-0001, Multidisciplinary Center for Earthquake Engineering Research, Buffalo, New York, pp. 439-453.

O'Rourke, M. and Nordberg, C., (1992), "Behavior of Buried Pipelines Subject to Permanent Ground Deformation," *Tenth World Conference on Earthquake Engineering*, Madrid, Spain, July 19-24, Vol. 9, pp. 5411-5416.

O'Rourke, M.J. and Ballantyne, D., (1992), *Observations on Water System and Pipeline Performance in the Limon Area of Costa Rica Due to the April 22, 1991 Earthquake*, Technical Report NCEER-92-0017, Multidisciplinary Center for Earthquake Engineering Research, Buffalo, New York.

O'Rourke, M.J. and Ayala, G., (1993), "Pipeline Damage Due to Wave Propagation," *Journal of Geotechnical Engineering*, ASCE, September, Vol. 119, No. 9, pp. 1490-1498.

O'Rourke, M.J. and Liu, X.J., (1994), "Failure Criterion for Buried Pipe Subjected to Longitudinal PGD: Benchmark Case History," *Proceedings of the Fifth U.S.-Japan Workshop on Earthquake Resistant Design for Lifeline Facilities and Countermeasures Against Soil Liquefaction*, Snowbird, Utah, Technical Report NCEER-94-0026, Multidisciplinary Center for Earthquake Engineering Research, Buffalo, New York, pp. 639-652.

O'Rourke, M.J., Liu, X.J., and Flores-Berrones, R., (1995), "Steel Pipe Wrinkling Due to Longitudinal Permanent Ground Deformation," *Journal of Transportation Engineering*, September/October, Vol. 121, No. 5, pp. 443-451.

O'Rourke, M. and Bouabid, J., (1996), "Analytical Damage Estimates for Concrete Pipelines," *Proceedings of Eleventh World Conference on Earthquake Engineering*, Acapulco, Mexico, June 23-28, No. 346.

O'Rourke, T.D. and Trautmann, C.H., (1980), *Analytical Modeling of Buried Pipeline Response to Permanent Earthquake Displacements*, Report No. 80-4, School of Civil Engineering and Environmental Engineering, Cornell University, Ithaca, New York, July.

O'Rourke, T.D. and Trautmann, C.H., (1981), "Earthquake Ground Rupture Effects on Jointed Pipe," *Proceedings of the Second Specialty Conference of the Technical Council on Lifeline Earthquake Engineering*, ASCE, August, pp. 65-80.

O'Rourke, T.D. and Tawfik, M.S., (1983), "Effects of Lateral Spreading on Buried Pipelines During the 1971 San Fernando Earthquake," *Earthquake Behavior and Safety of Oil and Gas Storage Facilities, Buried Pipelines and Equipment*, PVP-77, ASME, New York, June, pp. 124-132.

O'Rourke, T.D., Grigoriu, M.D., and Khater, M.M., (1985), "A State of the Art Review: Seismic Response of Buried Pipelines," *Decade of Progress in Pressure Vessel Technology*, (C. Sundararajan, Editor), ASME.

O'Rourke, T.D., (1988), "Critical Aspects of Soil-Pipeline Interaction for Large Ground Deformation," *Proceedings of the First Japan-U.S. Workshop on Liquefaction, Large Ground Deformation and Their Effects on Lifeline Facilities*, Tokyo, Japan, November, pp. 118-126.

O'Rourke, T.D. and Pease, J.W., (1992), "Large Ground Deformations and Their Effects on Lifelines: 1989 Loma Prieta Earthquake," *United States Case Studies of Liquefaction and Lifeline Performance During Past Earthquakes*, Technical Report NCEER-92-0002, Multidisciplinary Center for Earthquake Engineering Research, Buffalo, New York.

O'Rourke, T.D. and Palmer, M.C., (1994), "Earthquake Performance of Gas Transmission Pipelines," *Proceedings of the Fifth U.S.-Japan Workshop on Earthquake Resistant Design of Lifeline Facilities and Countermeasures Against Soil Liquefaction*, Snowbird, Utah, Technical Report NCEER-94-0026, Multidisciplinary Center for Earthquake Engineering Research, Buffalo, New York, pp. 679-702.

O'Rourke, T.D. and Meyersohn, W.D., Shiba, Y., and Chaudhuri, D., (1994), "Evaluation of Pile Response to Liquefaction-Induced Lateral Spread," *Proceedings of the Fifth U.S.-Japan Workshop on Earthquake Resistant Design of Lifeline Facilities and Countermeasures Against Soil Liquefaction*, Snowbird, Utah, Technical Report NCEER-94-0026, Multidisciplinary Center for Earthquake Engineering Research, Buffalo, New York, pp. 457-479.

O'Rourke, T.D., Gowdy, T.E., Stewart, H.E., and Pease, J.W., (1991), "Lifeline Performance and Ground Deformation in the Marina During 1989 Loma Prieta Earthquake," *Proceedings of the Third Japan-U.S. Workshop on Earthquake Resistant Design of Lifeline Facilities and Countermeasures for Soil Liquefaction*, San Francisco, California, NCEER-91-0001, Multidisciplinary Center for Earthquake Engineering Research, Buffalo, New York, pp. 129-146.

O'Rourke, T.D. and O'Rourke, M.J., (1995), "Pipeline Response to Permanent Ground Deformation: A Benchmark Case," *Proceedings of the Fourth U.S. Conference on Lifeline Earthquake Engineering*, Technical Council on Lifeline Earthquake Engineering, Monograph No. 6, ASCE, pp. 288-295.

Porter, K.A., Scawthorn, C., Honegger, D.G., O'Rourke, T.D., and Blackburn, F., (1991), "Performance of Water Supply Pipelines in Liquefied Soil," *Proceedings of the Fourth U.S.-Japan Workshop on Earthquake Disaster Prevention for Lifeline Systems*, Los Angeles, pp. 3-17.

Prior, J.C., (1935), Investigation of Bell and Spigot Joints in Cast Iron Water Pipes, Bulletin No. 87, *The Engineering Experiment Station*, Ohio State University Studies Engineering Series, Vol. IV, No. 1, January.

Ramberg, W. and Osgood, W., (1943), *Description of Stress-Strain Curves by Three Parameters*, Technical Note, No. 902, National Advisory Committee for Aeronautics, 28p.

Roth, B.L., O'Rourke, T.D., Dobry, R., and Pierce, C.E., (1990), "Lifeline Performance During the 1979 Imperial Valley Earthquake," *Proceedings of the Third Japan-U.S. Workshop on Earthquake Resistant Design of Lifeline Facilities and Countermeasures for Soil Liquefaction*, San Francisco, California, Technical Report NCEER-91-0001, Multidisciplinary Center for Earthquake Engineering Research, Buffalo, New York, pp. 191-207.

Rowe, R.K. and Davis, E.H., (1982a), "The Behavior of Anchor Plates in Clay," *Geotechnique*, Vol. 32, No. 1, pp. 9-23.

Rowe, R.K. and Davis, E.H., (1982b), "The Behavior of Anchor Plates in Sand," *Geotechnique*, Vol. 32, No. 1, pp. 25-41.

Sakurai, A. and Takahashi, T., (1969), "Dynamic Stress of Underground Pipelines During Earthquakes," *Proceedings of the Fourth World Conference on Earthquake Engineering*, Chilean Association on Seismology and Earthquake Engineering, Santiago, Chile, pp. 811-895.

Sato, K., Hamada, M., and Doi, M., (1994), "An Experimental Study of Effects of Laterally Flowing Ground on In-Ground Structures," *Proceedings of the Fifth U.S.-Japan Workshop on Liquefaction, Large Ground Deformation and Their Effects on Lifelines*, Snowbird, Utah, Technical Report NCEER-94-0026, Multidisciplinary Center for Earthquake Engineering Research, Buffalo, New York, pp. 405-414.

Sato, R. and Shinozuka, M., (1991), "GIS-Based Interactive and Graphic Computer System to Evaluate Seismic Risks on Water Delivery Networks," *Proceedings of the Third U.S. Conference on Lifeline Earthquake Engineering*, Technical Council on Lifeline Earthquake Engineering, Monograph No. 4, ASCE, pp. 651-660.

Schwab, F. and Knopff, L., (1977), "Fast Surface Waves and Free Mode Computations," *Seismology: Surface Waves and Earth Oscillation Methods in Computation*, Physics 11, Academic Press, New York, New York, pp. 87-180.

Seed, H.B., Idriss, F., and Arango, I., (1987), "Evaluation of Settlements in Sands Due to Earthquake Shaking," *Journal of Geotechnical Engineering*, August, Vol. 113, No. 8, pp. 861-877.

Shah, H. and Chu, S., (1974), "Seismic Analysis of Underground Structural Elements," *Journal of The Power Division*, ASCE, July, Vol. 100, No. PO1, pp. 53-62.

Shinozuka, M. and Koike, T., (1979), "Estimation of Structural Strains in Underground Lifeline Pipes," *Lifeline Earthquake Engineering - Buried Pipelines, Seismic Risk, and Instrumentation*, PVP-34, ASME, pp. 31-48.

Siegel, R.A., (1978), *STABL User Manual: West Lafayette, Indiana*, Purdue University, 104p.

Singhal, A.C., (1983), "Pull Out and Bending Experiments in Buried Pipes," *Earthquake Behavior and Safety of Oil and Gas Storage Facilities, Buried Pipelines and Equipment*, PVP-77, ASME, New York, June, pp. 294-303.

Singhal, A.C., (1984), "Nonlinear Behavior of Pipeline Joints," *Proceedings of the Eighth World Conference on Earthquake Engineering*, San Francisco, California, Vol. 7, pp. 207-214.

Steinbrugge, K.V. and Moran, D.F., (1954), "An Engineering Study of the Southern California Earthquake of July 21, 1952 and Its Aftershocks," *Bulletin of the Seismological Society of America*, Vol. 44, pp. 199-462.

Sun, S. and Shien, L., (1983), "Analysis of Seismic Damage to Buried Pipelines in Tangshan Earthquake," *Earthquake Behavior and Safety of Oil and Gas Storage Facilities, Buried Pipelines and Equipment*, PVP-77, ASME, New York, June, pp. 365-367.

Suzuki, H., (1988), "Damage to Buried Pipes Caused by Large Ground Displacement," *Proceedings of the First Japan-U.S. Workshop on Liquefaction, Large Ground Deformation and Their Effects on Lifeline Facilities*, Tokyo, Japan, pp. 127-132.

Suzuki, N., Arata, O., and Suzuki, I., (1988), "Subject to Liquefaction-Induced Permanent Ground Displacement," *Proceedings of First Japan-U.S. Workshop on Liquefaction, Large Ground Deformation and Their Effects on Lifeline Facilities*, Tokyo, Japan, pp. 155-162.

Suzuki, N., Kobayashi, T., Nakane, H., and Ishikawa, M., (1989), "Modeling of Permanent Ground Deformation for Buried Pipelines," *Proceedings of the Second U.S.-Japan Workshop on Liquefaction, Large Ground Deformation and Their Effects on Lifelines*, Buffalo, New York, Technical Report NCEER-89-0032, Multidisciplinary Center for Earthquake Engineering Research, Buffalo, New York, pp. 413-425.

Suzuki, N. and Masuda, N., (1991), "Idealization of Permanent Ground Movement and Strain Estimation of Buried Pipes," *Proceedings of the Third Japan-U.S. Workshop on Earthquake Resistant Design of Lifeline Facilities and Countermeasures for Soil Liquefaction*, San Francisco, California, NCEER Report Number 91-0001, Multidisciplinary Center for Earthquake Engineering Research, Buffalo, New York, pp. 455-469.

Takada, S., (1984), "Model Analysis and Experimental Study on Mechanical Behavior of Buried Ductile Iron Pipelines Subjected to Large Ground Deformations," *Proceedings of the Eighth World Conference on Earthquake Engineering*, San Francisco, Vol. VII, pp. 255-262.

Takada, S., Tanabe, K., Yamajyo, K., and Katagiri, S., (1987), "Liquefaction Analysis for Buried Pipelines," *Proceedings of the Third International Conference on Soil Dynamics and Earthquake Engineering.*

Takada, S. and Tanabe, K., (1988), "Estimation of Earthquake Induced Settlements for Lifeline Engineering," *Proceedings of the Ninth World Conference Earthquake Engineering,* August, Vol. VII, pp. 109-114.

Takada, S., (1991), *Lifeline Earthquake Engineering,* (In Japanese), 241p.

Tanabe, K., (1988), *Fundamental Study on Seismic Assessment and Design of Buried Pipelines Subjected to Ground Failure During Earthquake,* Doctoral Dissertation, Kobe University, 1988 (in Japanese).

Tawfik, M.S. and O'Rourke, T.D., (1985), "Load-Carrying Capacity of Welded Slip Joints," *Journal of Pressure Vessel Technology,* February, Vol. 107, pp. 37-43.

Tawfik, M.S. and O'Rourke, T.D., (1986), *Analysis of Pipelines Under Large Ground Deformations,* Geotechnical Engineering Report 86-1, School of Civil and Environmental Engineering, Cornell University, Ithaca, New York, March.

Thomas, H.O., (1978), "Discussion of Soil Restraint Against Horizontal Motion of Pipes," (by J. M. E. Audibert and K. J. Nyman), *Journal of the Geotechnical Engineering Division,* ASCE, September, Vol. 10, No. GT9, pp. 1214-1216.

Tokimatsu, K. and Seed, H.B., (1987), "Evaluation of Settlements in Sands Due to Earthquake Shaking," *Journal of Geotechnical Engineering,* ASCE, August, Vol. 113, No. 8, pp. 861-878.

Tomlinson, M.J., (1957), "The Adhesion of Piles Driven into Clay Soils," *Proceedings of the Fourth International Conference on Soil Mechanics and Foundation Engineering,* London, England, Vol. 2, pp. 66-71.

Towhata, I., Tokida, K., Tamari, Y., Matsumoto, H., and Yamada, K., (1991), "Prediction of Permanent Lateral Displacement of Liquefied Ground by Means of Variational Principle," *Proceedings of the Third Japan-U.S. Workshop on Earthquake Resistant Design of Lifeline Facilities and Countermeasures for Soil Liquefaction,* Technical Report NCEER-91-0001, Multidisciplinary Center for Earthquake Engineering Research, Buffalo, New York, pp. 237-252.

Trautmann, C.H. and O'Rourke, T.D., (1983), "Load-Displacement Characteristics of a Buried Pipe Affected by Permanent Earthquake Ground Movements" *Earthquake Behavior and Safety of Oil and Gas Storage Facilities, Buried Pipelines and Equipment,* PVP-77, ASME, New York, June, pp. 254-262.

Turner, W.G. and Youd, T.L., (1987), *National Map of Earthquake Hazard,* Final Report to the U.S. Geological Survey for Grant No. 14-08-001-G1187, Department of Civil Engineering, Brigham Young University.

Varnes, D.J., (1978), "Slope Movement Types and Processes," *Landslides Analysis and Control,* Special Report 176, Transportation Research Board, National Academy of Sciences, Washington, D.C., pp. 11-33.

Vesic, A.S., (1961), "Beams on Elastic Subgrade and the Winkler's Hypothesis," *Proceedings of the Fifth International Conference on Soil Mechanics and Foundation Engineering,* Vol. 1.

Vesic, A.S., (1971), "Breakout Resistance of Objects Embedded in Ocean Bottom," *Journal of the Soil Mechanics and Foundations Division*, ASCE, September, Vol. 97, No. SM9, pp. 1183-1206.

Wakamatsu, K. and Yoshida, N., (1994), "Ground Deformations and Their Effects on Structures in Midorigaoka District, Kushiro City, During the Kushiro-oki Earthquake of January 15, 1993," *Proceedings of the Fifth U.S.-Japan Workshop on Earthquake Resistant Design of Lifeline Facilities and Countermeasures Against Soil Liquefaction*, Snowbird, Utah, Technical Report NCEER-94-0026, Multidisciplinary Center for Earthquake Engineering Research, Buffalo, New York, pp. 41-62.

Waller, R. and Ramanathan, M., (1980), "Site Visit Report on Earthquake Damages to Water and Sewage Facilities El Centro, California, November 15, 1979," *Reconnaissance Report, Imperial Valley, California Earthquake, November 15, 1979*, EERI, February, pp. 97-106.

Wang, L.R. and O'Rourke, M., (1978), "Overview of Buried Pipelines Under Seismic Loading," *Journal of the Technical Councils*, ASCE, November, Vol. 104, No. TC1, pp. 121-130.

Wang, L.R.L., (1979), "Some Aspects of Seismic Resistant Design of Buried Pipelines," *Lifeline Earthquake Engineering - Buried Pipelines, Seismic Risk, and Instrumentation*, PVP-34, ASME, pp. 117-131.

Wang, L.R.L. and Yeh, Y., (1985), "A Refined Seismic Analysis and Design of Buried Pipeline for Fault Movement," *Journal of Earthquake Engineering and Structural Dynamics*, Vol. 13, pp. 75-96.

Wang, L.R.L., (1990), "Performance of Water Pipeline Systems from 1987 Whittier Narrows, California Earthquake," *Proceedings of the Fourth U.S. National Conference on Earthquake Engineering*, May, Vol. 1, pp. 965-974.

Wang, L.R.L., (1994), "Essence of Repair and Rehabilitation of Buried Lifeline Systems," *Proceedings of the Second China-Japan-U.S. Trilateral Symposium on Lifeline Earthquake Engineering*," Xi'an, China, April, pp. 247-254.

Wells, D.L. and Coppersmith, K.J., (1994), "New Empirical Relationships among Magnitude, Rupture Length, Rupture Width, Rupture Area, and Surface Displacement," *Bulletin of the Seismological Society of America*, August, Vol. 84, No. 4, pp. 974-1002.

Wilson, R.C. and Keefer, D.K., (1983), "Dynamic Analysis of a Slope Failure from the 6 August 1979 Coyote Lake, California Earthquake," *Bulletin of the Seismological Society of America*, Vol. 73, No. 3, pp. 863-877.

Wood, H.O., (1933), "Preliminary Report on the Long Beach Earthquake," *Bulletin of the Seismological Society of America*, April, Vol. 23, No. 2, pp. 43-56.

Wright, J. and Takada, S., (1978), "Earthquake Relative Motions for Lifelines," *Proceedings of the Fifth Japan Earthquake Engineering Symposium*, Tokyo.

Yasuda, S., Saito, K., and Suzuki, N., (1987), "Soil Spring Constant on Pipe in Liquefied Ground," *Proceedings of the 19th JSCE Conference on Earthquake Engineering*, pp. 189-192 (in Japanese).

Yasuda, S., Nagase, H., Kiku, H., and Uchida, Y., (1991), "A Simplified Procedure for the Analysis of the Permanent Ground Displacement," *Proceedings of the Third Japan-U.S. Workshop on Earthquake Resistant Design of Lifeline Facilities and Countermeasures for Soil Liquefaction*, San Francisco, California, Technical Report NCEER-91-0001, Multidisciplinary Center for Earthquake Engineering Research, Buffalo, New York, pp. 225-236.

Yeh, G., (1974), "Seismic Analysis of Slender Buried Beams," *Bulletin of the Seismological Society of America*, October, Vol. 64, No. 5, pp. 1551-1562.

Yeh, Y.H. and Wang, L.R.L., (1985), "Combined Effects of Soil Liquefaction and Ground Displacement to Buried Pipelines," *Proceedings of the 1985 Pressure Vessels and Piping Conference - Seismic Performance of Pipelines and Storage Tanks*, PVP-98-4, ASME, pp. 43-52.

Yoshida, T. and Uematsu, M., (1978), "Dynamic Behavior of a Pile in Liquefaction Sand," *Proceedings of the Fifth Japan Earthquake Engineering Symposium*, pp. 657-663 (in Japanese).

Youd, T.L. and Perkins, D.M., (1987), "Mapping of Liquefaction Severity Index," *Journal of Geotechnical Engineering*, ASCE, Vol. 113, No. 11, pp. 1374-1392.

Youd, T.L. and Garris, C.T., (1995), "Liquefaction-Induced Ground-Surface Disruption," *Journal of Geotechnical Engineering*, ASCE, November, Vol. 121, No. 11, pp. 805-809.

Zhang, B., Papageorgiou, A., and Tassoulas, J., (1995), "A Hybrid Numerical Technique, Combining the Finite Element and Boundary Element Methods, for Modeling Elasto-dynamic Scattering Problems," *Proceedings of the Tenth ASCE Engineering Mechanics Specialty Conference*, Boulder, Colorado, May 21-24, pp. 417-420.

Zhang B. and Papageorgiou, A., (1996), "Simulation of the Response of the Marina District Basin, San Francisco, California, to the 1989 Loma Prieta Earthquake," *Bulletin of the Seismological Society of America*, October, Vol. 86, No. 5, pp. 1382-1400.

AUTHOR INDEX

A

Akiyoshi, T., 219
Ando, H., 48-49
Arai, H., 217
Arango, I., 31-32
Arata, O., 87, 129, 115-117, 119-
    120, 122, 126, 137-139, 141-
    142, 146
Ariman, T., 64, 157-158, 163
Audibert, J.M.E., 85-86
Ayala, G., 4-6, 34, 72, 188-190

B

Ballantyne, D., 2, 61, 69, 70, 219
Barenberg, M.E., 4
Bartlett, S.F., 20, 22-26
Bazier, M.H., 22
Blackburn, F., 7-10
Bloom, M.C., 38
Boore, D.M., 35
Bouabid, J., 72, 204-207
Bozorgnia, Y., 35
Brockenbrough, R.L., 67-68
Bukovansky, M., 215

C

Cai, J., 143-146
Campbell, K.W., 35
Castro, G., 40-42, 44
Chaudhuri, D., 88
Chen, G., 143-144
Chow, A.W., 150, 153, 155-156,
    162-163, 166
Chu, S., 191-193, 196-197
Coppersmith, K.J., 14-15

D

Daniels, J.H., 63
Derbalian, G.K., 67

Dieckgrafe, R.E., 2
Dobry, R., 22, 38, 164
Doi, M., 88-89

E

Eguchi, R.T., 3-5, 8, 10-11
Eidinger, J.M., 5, 9
El Hmadi, K., 71, 73, 86, 169, 183-
    187, 189-190, 202-204, 209-
    210
Emoto, K., 23, 27-28, 94
Escalante, L.E., 2

F

Flores-Berrones, R., 36-37, 93-95,
    97-101, 170-171
Ford, D.B., 221
Fuchida, K., 219
Fujii, Y., 217

G

Garris, C.T., 213
Goodling, E.C., 191-193, 196-197
Gowdy, T.E., 30-31, 71
Greenwood, J.H., 215
Gregor, N.J., 37
Grigoriu, M.D., 1-2, 11-12, 35, 79

H

Hall, J.F., 46
Hall, W.J., 60-61, 150-152, 162
Hamada, M., 2, 8, 22-23, 27-29,
    88-89, 94
Hamura, A., 211-213
Hansen, W.R., 2
Hasegawa, H., 217
Haskell, N.A., 40
Heubach, W.F., 9
Hobbs, R.E., 63-64

Hojo, S., 207-209
Honegger, D.G., 5, 7-11
Hossain, I., 40-42, 44
Hou, Z., 11, 143-146
Howard, J.H., 62

I

Idriss, F., 31-32
Isenberg, J., 2, 219
Ishihara, K., 212-213
Ishikawa, M., 115-117, 120-121,
  137-138
Ishita, O., 47-49
Isoyama, R., 11, 22-23, 27-28, 94
Iwamoto, T., 199, 207-209
Iwatate, T., 217

J

Jibson, R.W., 16, 18-19
Joyner, W.B., 35

K

Kachadoorian, R., 46
Kamiyama, M., 36-37
Katagiri, S., 87, 143
Katayama, T., 3-4, 11
Keefer, D.K., 16, 18-19
Kennedy, R.P., 150, 153, 155-156,
  162-163, 166
Khater, M.M., 1, 2, 35, 79
Kiku, H., 22, 87
Kitaura, M., 87-88, 130-131, 137-
  138, 217
Knopff, L., 40
Kobayashi, T., 115-117, 120-121,
  137-138
Koike, T., 180, 182-183, 189, 194,
  196-197
Kondo, M., 87
Krathy, R.G., 72
Kubo, K., 3-4, 180
Kyriakides, S., 64

L

Lau, B., 5, 9
Lee, B.J., 64, 157-158, 163
Lee, D., 5, 9

Liu, X.J., 11, 49, 51, 54-55, 93, 95,
  97-102, 117, 120-128, 133,
  137-141, 143-146, 170-171

M

Mabey, M.A., 22
Maison, B., 5, 9
Major, G., 215
Manson, M., 2
Marek, P.J., 63
Markov, I., 11-12
Masuda, N., 25-26, 140, 147, 173
Matsumoto, H., 20, 22, 87
McCaffrey, M.A., 2, 149
McNorgan, J.D., 62
Meyersohn, W.D., 14, 16-17, 64-
  66, 88, 157-159, 180
Mijayima, M., 87-88, 130-131,
  137-138, 217
Moncarz, P.D., 67
Moran, D.F., 2

N

Nagase, H., 22, 87
Nakane, H., 115-117, 120-121,
  137-138
Newmark, N., 17, 44, 60-61, 150-
  152, 162, 180, 189
Nishio, N., 47-49, 56, 211-213
Nordberg, C., 92, 94, 171-173
Nozawa, Y., 217
Nyman, K.J., 85-86

O

O'Rourke, M.J., 2, 4-6, 34, 36-38,
  40-42, 60-61, 69-73, 85-86, 92-
  102, 116-117, 120-129, 169-
  173, 180, 183-190, 202-204,
  206-207, 209-210
O'Rourke, T.D., 1, 2, 7-12, 29-31,
  35, 60, 67-69, 71, 77-79, 81-
  82, 88, 96, 105, 113, 115, 117-
  119, 122, 129, 137-139, 149,
  157, 164, 175-177, 202
Ohtomo, K., 217
Orense, R., 22
Osgood, W., 60

P

Palmer, M.C., 2
Papageorgiou, A., 46-47
Pease, J.W., 2, 30-31, 71
Perkins, D.M., 23
Pierce, C.E., 164
Porter, K.A., 7-10
Prior, J.C., 71, 174

R

Ramanathan, M., 2
Ramberg, W., 60
Richardson, E., 219
Roth, B.L., 164

S

Saito, K., 87
Sakurai, A., 180-181
Salvadori, M.G., 72
Sasaki, Y., 87
Sase,T., 211-213
Sato, K., 88-89
Sato, N., 3-4
Sato, R., 11
Sato, S., 48-49
Scawthorn, C., 7-10
Schwab, F., 40
Seed, H.B., 31-32
Shah, H., 191-193, 196-197
Shiba, Y., 88
Shien, L, 2, 69
Shinozuka, M., 11, 180, 182-183,
    189, 194, 196-197
Shirinashihama, S., 219
Shyne, J.C., 67
Siegel, R.A., 16
Singhal, A.C., 71, 219
Steinbrugge, K.V., 2
Stewart, H.E., 30-31, 71
Sun, S., 2, 69
Suzuki, H., 114, 167
Suzuki, I., 87, 115-117, 119-120,
    122, 126, 129
Suzuki, N., 25-26, 87, 115-117,
    119-120, 122, 126, 129, 137-
    142, 146-147, 173

T

Takada, S., 31-32, 43, 87, 114,
    141, 143, 174-175
Takagi, N., 48-49
Takahashi, T., 180-181
Tamari, Y., 20, 22
Tanabe, K., 31-32, 87-88, 143
Tassoulas, J., 46
Tawfik, M.S., 29, 67-69, 105,
    113, 139, 157
Thomas, H.O., 86
Tohma, J., 217
Tokida, K., 20, 22
Tokimatsu, K., 31-32
Towhata, I., 20, 22
Towhata, T., 22
Trautmann, C.H., 77-78, 81-82,
    175-177, 202
Tsukamoto, K., 47-49, 56
Tsutsumi, T., 219
Turner, W.G., 23

U

Uchida, Y., 22, 87
Uematsu, M., 87
Ukaji, T., 47-49, 56

V

Varnes, D.J., 16
Vesic, A.S., 87

W

Wakai, N., 199
Wakamatsu, K., 2
Waller, R., 2
Wang, L.R.L., 2, 46, 85, 143,
    150, 156-157, 162-163, 180,
    200-201, 209-210, 219
Wells, D.L., 14-15
Williamson, R.A., 150, 153, 155-
    156, 162-163, 166
Wilson, R.C., 19
Wood, H.O., 2
Wright, J., 43

Y

Yamada, K., 20, 22
Yamaji, T., 199
Yamajyo, K., 87, 143
Yamamura, Y., 207-209
Yasuda, S., 22-23, 27-28, 87, 94
Yeh, G., 46, 180
Yeh, Y.H., 143, 150, 156-157, 162-163

Yew, C.H., 64
Yoshida, M., 217
Yoshida, N., 2
Yoshida, T., 87
Youd, T.L., 20, 22-26, 213
Yun, H.D., 64

Z

Zhang, B., 46-47

SUBJECT INDEX

A

Abrupt deformation
  of segmented pipelines, 170-171
American Society of Civil Engineers
  (ASCE) guidelines. *See* Technical
  Council on Lifeline Earthquake
  Engineering (TCLEE) Committee
  on Gas and Liquid Fuel Lifelines
Analytical models for pipe-soil
  interaction
  and strike slip faults, 150-157
  by Liu and O'Rourke, 133-137
  by M. O'Rourke, 131-133, 137
  by Miyajima and Kitaura, 130-
    131, 137
  comparison with case histories,
    139-141
  comparison with finite element
    analysis, 137-139
Anchorage, Alaska earthquake
  (1964)
  in empirical relations, 24
  pipeline damage, 2
Attenuation relations, 35-37
Axial pull-out
  of segmented pipelines, 70-72

B

Beam buckling, 62-66
Bell and spigot joints
  damage mechanisms, 70-72
  failure of, 59, 170-171
  joint efficiency, 67-69
Body waves
  definition, 33
  propagation velocity, 38
Break rates
  and serviceability index, 12
  of pipelines, 7

Buckling
  of beams, 62-66
  of continuous pipelines, 61-62
Buried pipelines
  and Anchorage, Alaska
    earthquake (1964), 2
  and Ecuador earthquake (1987), 3
  and liquefaction, 211-213
  and Mexico City earthquake
    (1985), 1, 72
  and permanent ground motion,
    xiii, 1
  and San Fernando earthquake
    (1971), 2, 3, 139, 149
  and San Francisco earthquake
    (1906), xiii-xiv, 1
  and wave propagation effects, 179
  as lifelines, xiii
  break rates, 7
  characteristics of, xiii, 75
  earthquake safety of, xiii
  empirical damage relations of, 3, 7
  seismic failure of, xiv, 59
  seismic performance in past
    earthquakes, 2-3
  use of, xiii
  wave propagation damage, 3-5
  wave propagation hazards, 2

C

Cast iron water pipes
  and permanent ground
    displacement, 7-9
  breakage, 8, 10-11, 71
  damage, 7
  damage prediction, 9
  empirical damage relations, 3-4

Circumferential flexural failure, 70-71, 73-75
Continuous pipelines
  analytical models for behavior
    due to fault movement
    Kennedy et al., 153-156, 162-163, 166
    Newmark and Hall, 150-152, 162-163
    Wang and Yeh, 156-157, 162-163
  and longitudinal permanent ground deformation, 91-97
  and Mexico City earthquake (1985), 189-190
  and transverse permanent ground deformation, 113-114
  case histories, 189-190
  comparison studies, 188-189
  Newmark's approach, 180-181
  O'Rourke and El Hamadi approach, 183-188
  Sakurai and Takahashi approach, 181-182
  Shinozuka and Koike approach, 182-183
  ASCE TCLEE guidelines
    Response to faulting, 166
  expansion joints, 102-105
  failure modes, 59-66, 113, 149-150
  finite element models for behavior due to fault movement
    Ariman and Lee, 157-158, 163
    Meyersohn, 157-159
  Models for
    elastic models, 92-96
    inelastic models, 97-99
    Ramberg Osgood models, 98-99
  with bends, 106-111, 191-197
  with elbows, 106-111
Costa Rica earthquake, (1991), 2
  buckling of continuous pipelines, 61
  damage mechanism for segmented pipelines, 70
Countermeasures to mitigate pipeline damage

isolation from damaging ground movement, 216-217
reducing ground movement, 217-218
routing and relocation, 215-216
using flexible materials and joints, 219-221
using high-strength materials, 218
Crushing of bell and spigot joints, 70-72

D
Damage mitigation techniques for pipelines
  flexible joints and materials, 219-221
  ground movement reduction, 217-218
  high strength materials, 218
  isolation techniques, 216-217
  relocation of pipelines, 215-216
  routing of pipelines, 215-216
Deformation of soil
  simulation of, 117-119
Distributed deformation
  of segmented pipelines, 168-170

E
Earth slides. See Landslides
Earth slump. See Landslides
Earthquakes
  Anchorage, Alaska (1964), 2. See also Anchorage, Alaska earthquake
  Borah Peak, Idaho, (1983), 24
  Coalinga, California (1983), 2, 179
    Costa Rica (1991) See Costa Rica earthquake
    Ecuador (1987), 3
    Guatemala, (1976), 2
  Imperial Valley, California (1979), 2. See also Imperial Valley earthquake
  Kern County, California (1952), 2
  Kushiro-Oki, Japan (1993), 2
  Loma Prieta, California (1989), 2. See also Loma Prieta earthquake

Long Beach, California (1933), 2
Managua (1972), 4, 8-9
Mexico City (1985), 1. *See also*
    Mexico City earthquake
New Madrid (1811-1812), 23-24
Nihonkai-Chubu, Japan (1983), 2.
    See also Nihonkai-Chubu
    earthquake
Niigata, Japan (1964), 2. *See also*
    Niigata earthquake
Northridge, California (1994), 2.
    See also Northridge earthquake
Off Miyagi Ken-Oki, Japan
    (1978), 48-49
San Fernando, California (1971),
    2-3. *See also* San Fernando
    earthquake
San Francisco (1906), xiii-xiv, 1,
    2. *See also* San Francisco
    earthquake
Superstition Hills, California
    (1987), 24
Tangshan, China (1976), 2. *See
    also* Tangshan earthquake
Whittier-Narrows, California
    (1987), 2
Empirical relations
    between fault displacement and
        moment magnitude, 14-15
    for amount of PGD due to
        liquefaction, 24-25
    for attenuation relations, 35
    for cast iron water pipes, 3-4, 7
    for faults, 14-15
    for liquefaction-induced seismic
        settlement, 31-32
    of buried pipelines, 3, 7
Expansion joints
    and continuous pipelines, 102-105
    and tensile stress, 105
    use of, 219-221

F
Failure mechanisms
    of continuous pipes, 149-150

Failure modes
    circumferential flexural failure,
        70-71, 73-75
    compression failure, 100-102
    of concrete cylinder pipelines, 72
    of continuous pipelines, 59-66
    of segmented pipelines, 69-75
    of welded slip joints, 67-69
    tensile failures, 59-66, 101-102
Fault crossing hazards
    analytical models for, 150-156
    finite element models for, 157-161
Fault movements
    types of, 13
Fault offset, 155, 160-163, 165
    and segmented pipelines, 174-177
Faults
    and permanent ground
        deformation, 13-15
    empirical relations, 14-15
Finite element analysis of pipe
    response
    by Kobayashi et al., 120
    by Liu and O'Rourke, 120-129
    by Suzuki et al., 119-120
    by T. O'Rourke, 117-119
    by Zhang et al., 46-47

G
Gas welded steel pipelines. *See*
    Steel pipelines
Ground curvature
    and seismic wave propagation,
        45-46
Ground displacement
    and pipe breaks, 7
Ground response
    numerical models, 47-53
    simplified models, 54-55
    subsurface conditions effects, 46-47
Ground strain
    and Miyagi Ken Oki, Japan
        earthquake (1978), 48-49
    and seismic wave propagation,
        44-46

and subsurface conditions, 46-47
comparison studies, 55-57
Newmark's procedure for
   estimating, 44
numerical models for, 47-53
simplified models for, 54-55

H
High strength materials
to mitigate pipe damage, 218

I
Imperial Valley, California earthquake, (1979), 2
attenuation relations, 35-36
buckling of continuous pipelines, 62
damage due to fault movement to
   continuous pipelines, 163-164
empirical relations, 24
propagation velocity, 38-39
Intensity, Modified Mercalli
and wave propagation pipe
   damage, 3, 5

J
Joints, crushing of, 70, 72
Joints, pipe
allowable angular offset, 74-75
Joints, rotation of, 73, 172-173

L
Landslides
and critical failure surface, 16-17
Newmark displacement, 18-19
types of, 16-17
Lateral spreading
analytical models for, 22
and permanent ground
   deformation, 21-25, 27
direction of, 20-21
effects on pipeline response, 20-21
empirical models for, 22-24
geometric characteristics, 20-21
numerical models for, 22
spatial extent of, 25-27
Liquefaction
and buried pipeline response,
   211-213

and Loma Prieta, California
   earthquake (1989), 211
Japanese Road Association
   liquefaction intensity, 89
Severity index, 23-25
Loma Prieta, California earthquake,
   (1989), 2
attenuation relations, 36-37
damage due to ground
   settlement, 30-31
damage due to liquefaction, 7-8
damage to segmented pipelines,
   71, 211
ground response analysis, 47
Longitudinal permanent ground
   deformation
and continuous pipelines, 91-107
   block pattern, 92-95, 97
   tensile failure, 101-102
   with elbow or bend, 106-109
   with expansion joints, 102-105
   wrinkling, 100-101
and elastic pipe models, 92-97
and inelastic pipe models, 97-99
ASCE guidelines for longitudinal
   movement, 79-81

M
Mexico City earthquake, (1985), 1
buckling of continuous pipelines, 61
damage due to joint crushing of
   segmented pipelines, 72
damage due to wave propagation
   to continuous pipelines, 179,
   188-190, 206-207
ground velocity, 34
Michoacan earthquake. *See* Mexico
   City earthquake

N
New Madrid earthquakes, (1811-
   1812)
empirical damage relations, 23-24
Nihonkai-Chubu, Japan earthquake,
   (1983), 2
cast iron pipe performance, 8
damage due to longitudinal PGD
   to segmented pipelines, 170

damage due to spatially
distributed PGD to segmented
pipelines, 173-174
effects of liquefaction, 27-28
empirical relations, 24
spatial extent of lateral spread
zone, 25-27
Niigata, Japan earthquake, (1964), 2, 4
damage due to liquefied soil to
continuous pipelines, 114,
140-141
damage due to spatially
distributed PGD to segmented
pipelines, 173-174
damage to segmented pipelines,
167
effects of lateral spread, 20
effects of liquefaction, 27-29
empirical relations, 24
spatial extent of lateral spread
zone, 25-27
Normal faults, 166
Northridge, California earthquake,
(1994), 2
buckling of continuous pipelines,
61
damage due to longitudinal PGD
to continuous pipelines, 95-97
damage due to wrinkling of
continuous pipelines, 100-101
effects of variable subsurface
conditions, 46
ground strain, 46
influence of elbows, 110-111
tensile failure of continuous
pipelines, 60

P
Peak ground acceleration
and pipe repairs, 4
attenuation relation for, 35
Permanent ground deformation
abrupt, 10-11, 113
and buried pipelines, 1, 7
and faults, 13
and local soil conditions, 6

and Niigata, Japan earthquake
(1964), 29
and San Fernando earthquake
(1971), 29
and segmented pipelines, 167
patterns of, 27
pipe break rate, 7, 10
spatially distributed, 7-9, 113-116
types of, 6, 13

R
R-waves. *See* surface waves
Repairs of pipelines, 218-221
Response of ground
numerical models for, 47-53
simplified models for, 54-57
Response of pipelines
and strike slip faults, 150-161
segmented pipelines, 167
Reverse faults, 13, 166
Rock topple. *See* Landslides
Rockfall. *See* Landslides
Rotation of joints, 73

S
S-waves. *See* Body waves
San Fernando, California earthquake, (1971), 2-3, 10
cast iron pipes, 8-9
damage due to expansion joints
to continuous pipelines, 105
damage due to fault movement
to continuous pipelines, 149-150
damage due to transverse PGD to
continuous pipelines, 113
damage due to welded slip joints
to continuous pipelines, 67
effect of liquefaction, 27, 29
effect of transverse PGD, 139
empirical relations, 24
ground acceleration records, 51
ground strains, 55
propagation velocity, 38-39, 43-44
transverse PGD patterns, 29
San Francisco Auxiliary Water
Supply System, 12

San Francisco, California earth-
    quake, (1906), xiii-xiv, 1-2, 7
  empirical relations, 24
  liquefaction, 20
Segmented pipelines
  and abrupt deformation, 170-171
  and distributed deformation,
    168-170
  and fault offsets, 174-177
  and longitudinal permanent
    ground deformation, 168-170
  and Niigata, Japan earthquake
    (1964), 167
  and Tangshan, China earthquake
    (1976), 69
  and transverse permanent ground
    deformation, 171-174
  and wave propagation effects, 199
  damage mechanisms for, 69-71
  failure modes, 69-75, 167, 171,
    199
    axial pull out, 71-72
    circumferential flexural failure, 73
  joint rotation, 73
  response to permanent ground
    deformation, 167
  with elbows and connections,
    207-209
Seismic hazards
  and buried pipelines, 1
  fault crossing hazards, 2, 150-157
Seismic vulnerability
  and fault offset, 11
  of pipelines, 8-9
Seismic wave propagation
  and damage to buried pipelines,
    2-5, 179
  and ground curvature, 45-46
  and ground strain, 44-45
  and subsurface conditions, 46-47
  and wavelength, 42-43
  effective propagation velocity,
    38-41
  idealized patterns of, 92-93
  types of, 33-34

Settlement
  and Loma Prieta, California
    earthquake (1989), 30
  causes of, 30
  evaluation of, 31-32
Soil-pipe interaction
  ASCE TCLEE guidelines for
    idealized elasto-plastic
    models, 77-78, 80-87
  axial movement, 85
  lateral movement in horizontal
    planes, 85-86
  liquified soils
    analytical approaches, 141-147
    soil springs, 87-88
    wrinkling of pipe walls, 100-101
  non-liquified soils
    comparison with case
      histories, 139-141
    comparison with finite element
      analysis, 137-139
    Kobayashi et al., 120-121, 137
    Liu and O'Rourke
      analytical, 133-137
      numerical, 120-129
    M. O'Rourke, 131-133, 137
    Miyajima and Kitaura, 130-
      131, 137
    Suzuki et al., 119-120, 137
    T. O'Rourke, 117-119, 137
  soil springs, 87-88
    equivalent stiffness of, 84-85
Spatially distributed permanent
    ground deformation, 6-9
  and Nihonkai-Chubu earthquake
    (1983), 173-174
  and Niigata, Japan earthquake
    (1964), 173-174
  and segmented pipelines, 171-174
Steel pipelines
  and empirical damage relations, 8-9
Straight segmented pipelines
  and compression, 204-207
  and tension, 199-204
  response to wave propagation, 210

Strike slip faults
  and response of continuous
    pipelines
      analytical models for, 150-157
      finite element models for, 157-
        162
Subsidence. *See* Settlement
Subsurface conditions
  and seismic wave propagation,
    46-47
  numerical models for, 47-53
  simplified models for, 54-55
Surface waves
  definition, 33
  propagation velocity, 39-41
System serviceability index, 12

T
Tangshan, China earthquake,
  (1976), 2
  damage to segmented pipelines, 69
Technical Council on Lifeline
  Earthquake Engineering (TCLEE)
    Committee on Gas and Liquid
    Fuel Lifelines Guidelines
  guidelines for modeling soil-pipe
    interaction, 77-87
  parameters in elastic pipe
    models, 101

finite element model parameters,
  120, 166
Tensile failures
  of continuous pipelines, 59-60
  of steel pipes, 101-102
Transverse permanent ground
  deformation, 88
  and continuous pipelines, 113-114
  and segmented pipelines, 171-174
  finite element analysis of, 117-129
  idealization of, 115-116
  localized abrupt patterns of, 147-
    148
  patterns, 29

W
Water supply system performance,
  11-12
Wave propagation. *See* Seismic
  wave propagation
Welded slip joints, 67-69
Welded steel pipelines. *See* steel
  pipelines
Wrinkling of pipe walls, 100-101
  ASCE guidelines for friction
    reduction factor, 101

C O N T R I B U T O R S

Dr. Michael J. O'Rourke

Professor
Department of Civil and Environmental Engineering
Rensselaer Polytechnic Institute
Troy, NY 12180-3590
orourm@rpi.edi

Dr. Xuejie Liu

Senior Engineer
Gulf Interstate Engineering Company
P.O. Box 56288
Houston, Texas 77256-6288
jacklliu@gie.com